WHEELS

A PICTORIAL HISTORY

When, in 1617, the Duke of Buckingham first hitched six horses to his coach, he was admiringly said to have developed "a masterly pride." By George II's time this pride had infected almost everyone who could afford it and, no doubt, many who could not. Four horses were enough to pull even those cumbersome shandrydans; the two extra animals were strictly for showing off, so were the servants perched on the rear axletree, and so, above all, was the "footman" who ran ahead brandishing a gold-headed staff, "to clear the way." This idea was copied from Indian rajahs. The man could keep ahead, because the team was trained to a rocking-horse gallop which was all action and no speed, like the gait of circus horses for bareback riding. The Duke of Queensberry once hired a new footman and equipped him with staff and expensive clothes, only to have the scurvy knave outrun the horses and get clean away with the outfit.

A Nobleman's Coach-and-Six in George II's Time, about 1730

WHEELS

A PICTORIAL HISTORY

WRITTEN AND ILLUSTRATED BY

Edwin Tunis

The Johns Hopkins University Press
Baltimore and London

ACKNOWLEDGMENTS

ONE OF THE REWARDS of building such a book as this one lies in the friendliness and generous kindness of the people who know most about the subject. First among them for this book is Paul Downing, a remarkable and charming gentleman who owns the best library on horse-drawn vehicles in the United States and is steeped to his ears in the whole lore of carriages and wagons. He is competent to build you a coach, or to drive one for you, and if necessary he can train the horses for the job; in fact, he has done all these things.

Colonel Downing feels that to justify his pursuit he should share the knowledge he has with writers and artists who are seeking information on the subject. Though he has in no way attempted to dictate what should go into this book, and in fact may disapprove mildly the inclusion of certain items, he has given days of his time and answered a steady flow of questions.

The friends and acquaintances who have searched their memories and albums and who have made valuable suggestions are too many to be named individually but I thank them collectively. Some people in official positions have obviously gone beyond the requirements of their jobs in being helpful. Of them I am grateful to Mr. Smith Hempstone Oliver, the Smithsonian Institution's Assistant Curator of Land Transportation; M. Max Terrier, Curator of the Musée de la Voiture, Compiègne, France; and Miss Elsa von Hohenhoff and Miss Margaret Jacobs of the Pratt Library in Baltimore.

My wife gets special thanks, not only for the mass of typing she has done, but also as the critic who saw to it that the wheels were headed right and the horses were of the proper size for the vehicles they pulled.

E. T.

Long Last
September 20, 1954

Copyright © 1955 by Edwin Tunis
All rights reserved.
Printed in the United States of America on acid-free paper

Originally published in 1955 by Thomas Y. Crowell Company
Johns Hopkins Paperbacks edition, 2002
2   4   6   8   9   7   5   3   1

Design and Typography Jos. Trautwein

The Johns Hopkins University Press
2715 North Charles Street
Baltimore, Maryland 21218-4363
www.press.jhu.edu

Library of Congress Cataloging-in-Publication Data
Tunis, Edwin, 1897–1973.
Wheels : a pictorial history / written and illustrated by Edwin Tunis.—
Johns Hopkins Paperbacks ed.
p.   cm.
Originally published: New York : Thomas Y. Crowell, 1955.
ISBN 0-8018-6929-3 (pbk. : alk. paper)
1. Wheels—History—juvenile literature.
2. Carriage and wagon making—History—juvenile literature. I. Title.

TS2025 .T86 2002
629.04´9—dc21   2002016043

A catalog record for this book is available from the British Library.

Baby cart, early nineteenth century

for Lib
this book
and all else
besides

ILLUSTRATIONS

FOREWORD

To simplify a complex subject it's necessary to leave some things out. Unfortunately when the edges of the omissions are drawn together, the result is an appearance of completeness which can be misleading. In this outline of the history of vehicles it has been necessary to turn away from much that would demand to be included in a full-scale study of the subject. No such opus now exists and for research purposes it is needed. Let it be said, however, that when it is written it will run to several volumes and it may be somewhat dull going for the general reader.

This book is for that general reader. With some allowance for human error it is as accurate as available information can make it. It aims to show what kind of conveyances people used in the past and how they were affected by the conditions of their times. Many of them are ingenious; some are strange; and a few are downright comical. Vehicles have reflected the aspirations of man and they have also reflected his preposterous vanities.

Much more space is given here to wagons and carriages than to automobiles. That is partly because this is a chronicle and automotive vehicles haven't a very long history. It is also because the history of automobiles is available in many books, while there has been no consistent and accurate account of horse-drawn vehicles published in the United States since 1878. A couple of centuries from now, when people are riding in some as-yet-inconceivable kind of conveyance, it may be easy to see the motorcar in its proper perspective against all that has gone before it. That's difficult to do today when this new self-propelled thing is such a dominant factor in our lives.

Wheels on rails and wheels in stationary machinery are no part of this book. It keeps to wheels rolling on the earth. It begins before there were any wheels and follows the development of vehicles up to yesterday morning.

E. T.

Colonial wagon jack

WHEELS

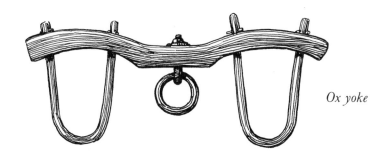

Ox yoke

PERHAPS IT ISN'T TRUE THAT if man could have been made in more than one piece, he would have had wheels instead of feet. It is certain, however, that nothing that works like a wheel has ever been discovered in nature. Man made it up completely, out of his head. Fire is his greatest discovery; the wheel his greatest invention.

The invention of the wheel has been credited to the Elamites because their sculptures are the earliest to portray it. Though there is a possibility that it may be older than their civilization, it almost certainly did originate in their neighborhood and nowhere in the world has the wheel ever been used except by people who came directly or indirectly into contact with the Tigris-Euphrates Valley. At Kish, in this area, were found the oldest actual wheels which have ever been discovered; they are estimated to be 5,000 years old.

The ancestors of European, of most Asian, and of some African peoples have used wheels for thousands of years, but over the rest of the earth they were unknown until modern times. Civilization runs on wheels; whenever it has reached a high point there has been a great increase in their use, particularly for carrying people from place to place.

Pack bullock, prehistoric

Pawnee Indian travois

Primitive sledge

Man civilized himself by finding easy ways to do things. Often that meant letting somebody else do the work. Cattle were tamed because it was easier to keep them penned up at home than to hunt them across the country. Presently man realized that his ox could carry more than he could. A horse could carry a burden faster than an ox, but it's likely that horses weren't tamed until many centuries after cattle were.

Nobody thought of a vehicle at first; the man simply took the load off his own back and put it on the back of his bullock. However, he soon found that an ox could drag even more weight than he could tote. Dragging usually required some kind of carrier. We can only guess what kind, but we can base one of our guesses on the way a much later primitive people solved the same problem.

Until the Spanish brought horses to America the Indians had never seen them, but they took to them at once and handled the dragging problem with two long poles, lashed together across a horse's back at one end and trailing on the ground at the other. Some Frenchman named this rig a *travois*. Actually the travois was not specially invented for the horse; it had already been used by the Indians as a dog-drawn vehicle.

Another reasonable guess, made long ago, is that the first vehicle may have been a sledge made from a naturally forked tree limb. The branches would serve as runners, and they would have two or three sticks lashed across them and would be hitched to an ox with a rawhide thong.

It's quite certain that runners came before wheels. Very old Egyptian wall paintings show sledges being used to carry mummies to tombs. This was long after wheels had been invented, but the Egyptians kept up the custom of their ancestors, thinking it disrespectful to transport the dead on wagons.

Long before the oldest records, some genius thought of moving a heavily loaded sledge on rollers, placing them on the ground ahead of the load, picking them up from behind after the sledge had passed over them, and rushing them around to the front again. Heavy machinery is still moved that way every day.

Egyptian sledge-hearse, before 2000 B.C.

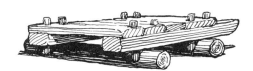

Sledge on captive rollers

Sledge on free rollers

Then it probably occurred to somebody to make life longer for the oxen by using rollers for light loads too. Pins put through the runners would hold the rollers captive and still allow them to turn.

Grooved roller

This next is pure guesswork: the rollers, turning under the sledge, would wear notches in the runners and, at the same time, the runners would groove the rollers. For a while this was good; it kept the rollers from creeping out endways. But as the grooves wore deeper, the middle part of a roller would begin to rub on the bottom of the sledge. The cure for this would be to make the roller thinner between the runners. As soon as this was done, the roller wasn't a roller any more; it was two wheels connected by an axletree, and the sledge became a wagon or, perhaps more likely, a cart.

There was a catch in having an axletree that turned with its wheels: both wheels were forced to rotate at the same speed and, for going around corners, that's bad, in fact, the wheels stop turning entirely and are simply skidded around the turn. So our ancient ancestors had only to try this out to learn that it was bad, and eventually the tribal thinker came up with the answer: wheels that could turn while their axles were held still. Then, at a corner, the inside wheel could creep, or even turn backward, while the outside one picked up speed and hustled around to where they could both start off in the new direction together.

Those first wheels were circular chunks cut, or perhaps burnt, from the ends of tree trunks, with a hole made in or near the middle for the axle to go through. The end of the axle stuck out beyond the wheel and had a pin through it to keep the wheel from coming off. Later this was called a linchpin. Tree-trunk wheels can still be found. The Romans used them on farm carts and called them *tympani*, because they were shaped like drums.

When Julius Caesar landed in Britain in

Solid "wheels" with rotating axle

Solid wheels on fixed axle

Early British war chariot, 55 B.C.

55 B.C., he found the blue-painted natives using drum-wheels on their war chariots. One of the chariots was taken to Rome as a curiosity. Of course, this was a very long time after the earliest wheels of this kind were made.

The aboriginal Britons felt that the safest place to travel was along the crest of a ridge, where you could see what was going on around you. The grass now grows with a different texture and color where their packed trails went, and the "green roads" are still clearly visible snaking along the tops of the downs.

Simple drum-wheels are easily split because of the way the grain of the wood runs in them. To overcome the splitting, solid wheels came to be built up of thick planks pinned together, the grain running at a right angle to the axle instead of parallel to it. There are plenty of such wheels creak-over the world right now. The disc-wheel of an automobile is an up-dated drum-wheel.

Solid wheels are very heavy and they almost never run true. They wobble and, as they usually have no tires, they wear unevenly and get pretty lumpy. By the way, don't let anybody tell you that square wheels are used in Tibet! Except for prayer wheels no wheels are used there at all, and as far as is known, nobody, anywhere, ever tried to use square ones, except perhaps as a feeble joke.

At some time before the beginning of recorded history, someone learned to make a wheel with a hub, spokes, and fellies. (A felly, or felloe, is one section of the rim of a wheel.) The first such wheels must have been as clumsy and crude as those on the Persian cart opposite, but they still were lighter and stronger than any kind of solid wheel. The fellies of these Persian wheels are not bent to shape, as later ones were; they are hewn from straight timber and labor was saved by leaving the inside unshaped.

At first and for a long time, hubs, axles, and linchpins were made entirely of wood. Naturally they didn't wear very well, and their creaking and squeaking was horrible to hear, so the axle was sheathed with metal and the hub was lined with it. Probably copper or bronze was used at first.

The first carts were used for hauling. The Elamite vehicle is a reconstruction of one of the first recorded passenger chariots, probably used for fighting. Its wheels were made in three pieces, clamped together with

Mexican cart with built-up solid wheels

copper, and they apparently had copper hubcaps and copper tires. This last is remarkable because centuries later the Assyrians and Egyptians were using wooden tires. Some of the neighbors of Elam stood up in their chariots, but the "body" of this vehicle seems to imitate the shape of an animal used for riding and the driver probably mounted accordingly. One may guess that it looked to the Elamites like a camel, since the horses and asses of the time are believed to have been too small to be ridden.

Horses probably hadn't been broken to harness; there's little question that the ancients were deathly afraid of them. The draft animals of this chariot are wild asses, called onagers. You'll notice that the harder they pulled the more likely they were to choke themselves. Though the collar was widened later and the yoke was moved farther back, it took a couple of thousand years for men to learn that an animal can't do his best pulling with his neck! It's still tried. Notice where the oxen pulling the Mexican cart have their yoke. As to stopping, it would seem that, with no bellyband, any holding-back effort would put the collar up around the asses' ears. Perhaps the bellyband was left out of early representations as a matter of artistic license; none of the oldest ones show it, so it has been left out here.

At about the time horses began to be used, the Assyrian sculptures show chariots with spoked wheels. These wheels had rather narrow fellies and thick wooden tires, clamped on; presumably they could be changed when they wore out. Still later chariots, like that of Sennacherib on the next page, had the face of the tire studded with nailheads to make it last longer. This chariot was the end product of centuries of slow development and was used as a carriage. It had high wheels to make it

Persian cart with crude spoked wheels, about 1870

Elamite chariot, about 2500 B.C.

ride easier and an umbrella to protect the king from the sun. The pole was much straighter than earlier Chaldean ones and hence stronger; it arched over the horses' backs almost as much as the pole of the Elamite chariot did and had to have a kind of forestay to strengthen it.

The Egyptians had the finest horses in the ancient world. If you wanted to pay your girl a compliment, you told her she looked like an Egyptian horse! The chariots of Egypt were handsome, too, very lightly built, perhaps because wood was scarce on the Nile, and better designed than those of Assyria. Sennacherib stood in his chariot well ahead of the wheels, so that the horses carried part of his weight; the Egyptian stood right over his axletree, with his whole weight supported by the wheels.

Both Egyptians and Assyrians guided their horses with reins attached to a bit,

Carriage-chariot of Sennacherib, King of Assyria, about 720 B.C.

which was held in the horse's mouth by a leather bridle, or headstall. Only the details of bits and bridles have changed, the essential parts are the same today as they were along the Nile. Sometimes the bits as well as the bridle were leather; more often they were wood, bone, or metal. By Roman times quite modern-looking bronze bits were in use and are still to be seen in museums.

The dry air of Egyptian tombs has preserved a few fragments of actual chariots, among them parts of a wheel, so that we know exactly how it was made. The hub was very long and slender; the spokes light and nicely shaped. The fellies, one to a spoke, were bent to shape and joined with a long lap, so that the inserted end of the spoke could hold them together. The wooden tires were cut to shape in sections and lashed to the wheel with rawhide passed through slots cut in the sections for

the purpose, so that the lashing wouldn't be worn through by contact with the ground. Egyptian war-chariot wheels had six spokes, the carriage chariot's only four. The chariot was pulled by its pole, which was supported by a yoke across the horses' backs, its ends being attached to saddle-pads held in place by girths around the horses' bellies.

The *harmanaxa* is drawn to fit Herodotus' description, but the ornamental details shouldn't be taken too literally. The Persians used their harmanaxa as a carriage. It's the earliest four-wheeled vehicle known to have been so used, and as late as 1919 the only means of travel between two Persian railheads was in a wagon almost exactly like the one in the drawing, less ornate but curtained in the same way and utterly innocent of springs. The Greeks gave one of their early kings credit for inventing the wagon, but the Chaldeans, who

Egyptian hunting chariot, about 1500 B.C.

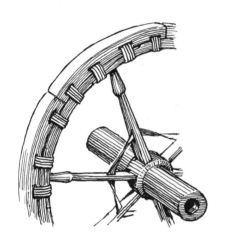

Part of an Egyptian wheel

Assyrian bridle (the bit was probably bone)

Persian harmanaxa *or women's wagon, about* 450 B.C.

used four-wheeled war chariots before 2000 B.C., beat him to it.

At first harmanaxas were used only by women, but in the later days of the Persian Empire elegant young men rode in them, to the deep disgust of their tougher elders. This wagon had a "box" body and running-gear something like that of the Roman farm wagon. There was a canopy top and curtains to keep out the sun and give privacy to the passengers. Its horses were yoked to a pole. The front and rear axletrees were connected by a perch, and there had to be a kingbolt to permit turning corners. (The perch is simply a pole rigidly fixed to the rear axletree and swiveled to the middle of the front one by a single pin, which is the kingbolt.) The harmanaxa could have had iron tires but, if so, they were riveted to the rim in sections. The skill of wheelwrights was not then, or for a long time afterward, equal to applying a continuous iron band.

The two right-hand wheels are left off the drawing of the Roman farm wagon running-gear so that its construction can be seen better. For the same reason the tire is omitted from the left rear wheel. In later days more iron and less wood was used for wagons and better ways were found to put them together, but, as with the bridle, the principle persists and the carriage of a modern farm wagon isn't too unlike this one. Iron was used on this wagon at the points where it had to be strongest. It wasn't always available, however. On page 67 there's a drawing of an American Red River cart, which was built by Western settlers without using any metal whatever.

In addition to wagons and carts for hauling, the Greeks used many vehicles for travel and convenience; most of these, but not all, were forms of the chariot. The cart opposite is supposed to have been for long journeys, though it pays no attention to comfort and little to safety. There are two odd things about it: first, its spokes do not radiate from the hub, as spokes regularly do, and second, the horses were driven without reins. Both of these things seem to have been a little unusual, even in ancient Greece. The driver sat on the bottom of the cart and guided his horses in some mysterious way with a long, hooked stick. In his left hand was a sharp goad to urge them on. It has been suggested that this way of driving may have been a kind of circus trick. Certainly the driver had no way of stopping his animals except by the strength of his personality! Anyway, this is how it shows on an ancient vase. The poor horses were still being choked by their effort to pull. The Greeks seem never to have used any more harness than yoke, bellyband, collar, bridle, and, usually, reins.

Reconstruction of the running-gear, or carriage, of an ancient Roman farm wagon

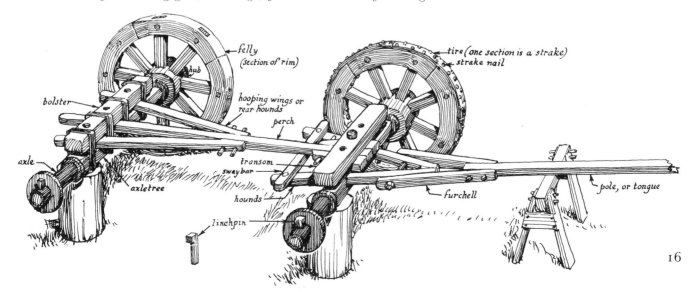

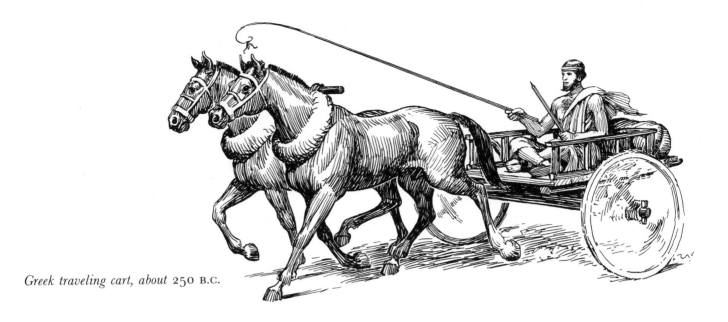

Greek traveling cart, about 250 B.C.

The simplest Greek chariot was a mere skeleton called a *diphron*. It had very small wheels and was used with two horses for racing. As with nearly all Greek vehicles, the axletrees were very long to avoid upsetting; they were usually made of beechwood but some were of iron. The front and sides were fairly high and the top rail was sometimes a hoop which completely enclosed the driver to hold him in and to give him support.

In Homeric times the Greeks used chariots in war. They abandoned them later and chariots were used as carriages only. The diphron has been miscalled a war chariot by several authorities.

The *biga* on the next page was a kind of chariot. It was used as a carriage for a lady who did her own driving. It's called a biga because it was used with two horses; any two-horse chariot is a biga. One with three horses is a *triga,* four a *quadriga.*

Greek carriages of this kind usually had wheels and axles of iron or bronze. There was a small seat, often not much more than a crossbar on a post, and the effort of staying on it must have given the ladies lots of exercise. Some dashing females also drove quadrigas; these, the chariots that is, were shaped like the man's quadriga, but with a wider body and a seat for two.

A Greek gentleman went about in a one-man quadriga. It had neat handles for aid in mounting and usually had a board across the tops of the two handles to serve as a seat. The Greeks traveled a lot around their own country; their very wide-gauge wheel ruts are to be found even in the most remote mountain passes.

The two outside horses in a quadriga weren't able to do much pulling. The yoke rested on the backs of the inside horses, the other two were merely hitched to its ends by a leather strap. It seems the only way the vehicle could have been held back, going downhill, would be for the inside horses to throw their heads up, taking the forward pressure of the yoke against the backs of their necks. Some chariots had a rope or strap leading from the body to the yoke presumably to equalize the pulling strain.

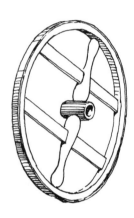

The wheel of the traveling cart

Greek racing chariot, about 250 B.C.

Greek lady's light biga, *about* 250 B.C.

Greek gentleman's quadriga, *about* 250 B.C.

Roman letica, *or litter,* 54 A.D.

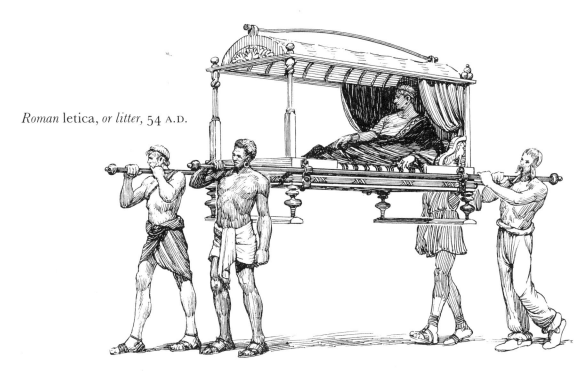

The Romans used chariots as carriages, though they did not use them in war. As Rome became powerful and captured many slaves, the usual conveyance in the city was a litter carried on the shoulders of four men. Litters became so numerous that they created traffic jams. The Emperor Claudius had the first covered litter, like the one in the drawing, in 54 A.D. In later Roman days litters completely enclosed by leather covers were used in bad weather and for privacy when it seemed to be indicated.

By this time a variety of wheeled vehicles were in use in Rome, but the *currus,* or standard chariot, remained, to the end of the Empire, the correct equipage for a Roman gentleman on business in the city. It was impressive but not very comfortable. There was no seat; a driver handled the horses and the passenger stood on the tailboard and struck an imposing attitude. It seems a little funny to us, especially in view of all this dignity, but sometimes the Romans hitched *mules* to their chariots.

The currus was pulled by a yoke as chariots had always been, but with a girth around the belly of the horse to give him a better chance to use his strength than the

Roman currus, *or town chariot,* 100 B.C. *to* 400 A.D.

yoke alone would give him. The yoke, or *jugum,* was a symbol of subjection to the Romans; they made their conquered enemies walk under one. Our word *subjugation* means, literally, "bringing under the yoke." The Roman axletree was shorter than the one the Greeks used, because Roman roads were better and there was less danger of being upset. City streets were paved and where sections of pavement have been preserved, as at Pompeii, deep ruts are found worn in the stone. These are usually said to have been made by chariot wheels, and no doubt they contributed, but there were other Roman carriages.

Going to the races was more than sport

Roman racing chariot, about 200 A.D.

in Rome; it was a privilege of citizenship and even a duty. The cost of the race meetings was borne by the state, and every citizen who possibly could be there was in the Circus Maximus to cheer. Once in, none might leave until all the courses were run, and heavy punishment fell upon any bored spectator who went to sleep.

As many as twenty-four matches were run in one day. Races were held in connection with eight public festivals, usually on the last day of the celebration. Each match was between four chariots identified by colors: green, red, blue, white. The fans favored a certain color to win, not a certain team or charioteer, though some of the drivers became public idols. At first the chariots were driven by slaves or freedmen but, in later years, noble youths and sometimes senators drove in the Circus.

The chariots were light and very narrow, with extremely low wheels, quite different from those which have been used for "racing" in modern circuses. The charioteer, wearing wrapped leather leggings and gloves, stood up in his cart; he seems to have wrapped the reins around his waist so that at need they could steady him. The four horses, too, wore wrapped leggings and leather masks and from the breast strap of the off-side horse hung a bronze bell, to add to the noise. (With horses, the off side

is the right. The horse's left is the near side.)

A race was seven times around. To mark each lap, one of seven egg-shaped counters was removed from the stone barrier around which the race was run. At the ends of the barrier were two stone pylons, each of them looking like three modern artillery shells upright, to mark the turns. By modern standards the course was very short and the turns sharp; winning depended more on tactics than on speed. The tactics weren't always subtle. The inside chariot ran much the shortest race; to compensate for this, the start was made from an oblique line.

Roman horse-yoke. The tip of the pole was secured in the ring and the ends of the yoke were strapped to the horses' backs

Roman roads largely disregarded hills and went where they were going in an absolutely straight line. The important ones were paved with stone and, at intervals of ten miles or so, post houses were maintained along them. Some post houses were police stations or military garrisons, and some had inns attached, but all provided fresh horses for travelers.

Of the many and varied vehicles on such

Roman post chaise, called a birotum, *about* 300 A.D.

Roman mail cart, or cisium

roads, not a few were public conveyances. One of these was the state-owned *birotum*, the direct ancestor of "the wonderful one-hoss shay" and the gig. Much used by government couriers, it could also be hired by private citizens and was the Roman version of the "drive-it-yourself." Equipped with a leather seat and a baggage rack, it made a handy carriage for a traveler who wasn't in too much hurry. The legal limit of baggage on a birotum was six hundred pounds, and the gig itself wasn't light, so it was not to be expected that its one horse could pull it (and a solid Roman citizen) very fast. However, fresh horses were constantly available and a man could make thirty or forty miles a day in a birotum. When a vehicle is pulled by one horse, he is usually put between two poles called shafts. The birotum's horse did all his pulling by the shafts; he had no traces to pull by.

People in haste could travel as passengers in a *cisium* with the mail. These were light carts in use at least as early as 55 B.C. They had regular drivers and kept a schedule whether anyone traveled in them or not. They too changed horses frequently. The drivers seem to have been hard characters. They held casual races on the road, to the horror of their passengers, and they were often punished for wild driving. Because they carried mail, these carts made exceptionally good time, about fifty-six miles in ten hours. The bodies of some of them are said to have been suspended on straps to make them ride easier and some had large wheels for the same purpose, but no pictures of these seem to have survived.

During the later Empire the *rheda* was the commonest wheeled vehicle in the streets of Rome. Usually a private carriage, though sometimes for hire as a hack, the rheda was a rather elaborate open cart with seats for six passengers.

Inside the city nearly all rhedas were pulled by oxen and progressed so slowly and steadily that Cicero could write letters while riding in one. Sometimes, especially in the country near Rome, mules were hitched to rhedas, so that they moved a little faster.

For long-distance travel a horse-drawn, post rheda which was covered, curtained, and cushioned was in use. There seem to be no pictures of these, but they had four

Roman rheda, *about* 300 A.D.

instead of two wheels, and Julius Caesar traveled ninety-five miles a day in one of them. The names of many Roman vehicles are known without anyone having an accurate idea of what they looked like. Others are pictured on Roman sculptures, coins, and wall paintings. Some of these in common use are shown here.

Carrus. *An ox or mule cart used for general hauling and as army baggage cart and ambulance! The shape of the body varied with use*

Monachus. *This very light one-horse or one-mule cart was used by Roman ladies for driving informally in the city*

Plaustrum. *The standard farm ox cart, having solid wheels reinforced, driven by a trained dog*

Carrus clabularius. *This was used as a dray for general hauling*

Carpentum. *Preferred to the litter because two could ride, the carpentum was a woman's vehicle*

Wheels were used for freight-hauling in China from very ancient times, but they were seldom put on carriages. The emperor's cart opposite was an exception; it seems that at one time only royalty was permitted to ride on wheels. It is possible that this cart was used as a fighting chariot. Lesser folk who could afford them used boxlike palanquins carried by coolies and were unquestionably much more comfortable in them than was the emperor in his cart.

Carruca. *Rome was dying when this ridiculous thing appeared. Its one purpose was to show off its occupant*

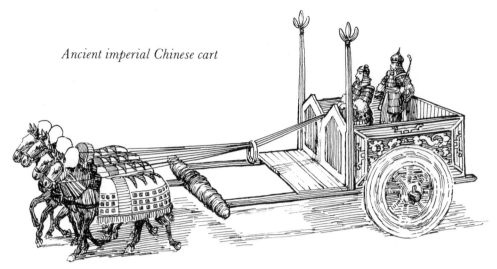

Ancient imperial Chinese cart

The Chinese artist who first drew this vehicle didn't bother to show what kept the long "outrigger" off the ground. Something had to hold it up, so a pole and yoke have been added here to the drawing. Probably in no other vehicle were the horses ever hitched so far from the wheels! These horses in the old drawing are absurdly small, no bigger than large dogs. Here they have been drawn about the size of a present-day Mongolian pony; even the prehistoric horse, the *tarpan,* was four feet high!

The narrow tires of the Peking cart have rutted Chinese roads for centuries and are likely to go on rutting them for a long time to come. A donkey pulls it, and it is used with various bodies for all kinds of hauling. Hooded carts like the one on the next page are carriages, but their use for passengers is dying out. As with the Roman birotum, the donkey is not hitched to the vehicle in any way except by the shaft-ends.

Perhaps the most universal Chinese vehicle is the wheelbarrow. Enormous loads are hauled on it. Chinese barrows have large wheels placed directly under the burden so as to carry most of its weight. The man at the handles usually has help; his wife or his donkey pulls on a rope ahead. A sail is occasionally rigged on a wheelbarrow when the wind is just right.

In the Gobi Desert the rich travel by camel cart, like the one on the next page: swinging from side to side, rocking and bumping twenty-five miles a day; a passenger with fewer layers of padded clothing than a Mongol wears would be bruised to a pulp. Such felt-hooded carts have been used in the Gobi since before 1200 A.D.

With the wheels set far back, as they are, the poor camel carries an unfair share of the cart's weight, in addition to which he usually has a burden loaded on his back. The ends of the shafts are merely hung from the saddle with ropes, and the camel must yank the cart along by them as best he can.

The Japanese use some many-spoked bullock carts on their farms, but passengers usually ride in a jinrikisha. The one on the next page was drawn from a photograph taken about 1890. The modern adaptation of the jinrikisha has rubber tires and wire wheels and is propelled from the rear by

Chinese wheelbarrow

Japanese jinrikisha, about 1890

Peking cart

pedals. It is really a tricycle and called a trisha. In a sense all 'rikshas are modern. They were introduced into Japan about 1870 by missionaries, who presumably invented them or copied them from the eighteenth-century brouette. (See page 42.) Before 1870 the Japanese rode in lacquered palanquins hung from a single pole and carried by two or more men.

In the days of pulling, or rather pushing from the front, a 'riksha boy could trot about seven miles an hour and was the possessor of a high-class set of leg muscles. The use of the jinrikisha has spread to other Oriental countries where manpower is plentiful and cheap. According to the book, the word in Japanese means quite literally, syllable by syllable, man-powered vehicle.

As with China, it is difficult to say when the vehicles of India started. They have been used for hundreds of years and they are still used. Small horses pull some of them, but zebus are more frequently seen. When the English first came to India their ears were tortured by the universal shriek-

ing of ungreased axles. There was then no mineral grease, and the Hindus recoiled with religious horror at the idea of using animal fat. After many words, a poor compromise was reached with olive oil.

There are railroads and buses in India now, but formerly journeys of hundreds of miles were made by bullock cart, plodding along by day and resting by the roadside at night. Speed is a Western invention. The driver rode the tripod pole. If there were women in the cart, they were kept hidden behind the curtains, regardless of the heat.

It might almost be true to say that Indian farmers have always used the *gujerat* cart. It runs in ruts. The ruts are the roads and are kept in reasonable repair. The carts are built to fit the ruts exactly and the width of the yoke forces the oxen to walk in the ruts, also.

The gujerats are strongly made, with mortised joints held together by wooden pins. It is not unusual for one to be loaded with nearly a ton of material. The double pole forms a long triangle, its apex at the

Mongolian camel cart

24

Hindu hecca *and passenger bullock cart* (ruth)

yoke, its base the axletree. This makes turning the cart unusually easy and is a great help in getting it out of the ruts when necessary. The wooden wheels are built with fellies and the double spokes which are characteristic of India. The axles are iron, and the hubs have iron linings or boxes.

The digression to give some notion of Oriental vehicles has carried this chronicle far ahead of itself in time. It returns now to the slow development of road transport in Europe.

When Rome collapsed, the Dark Ages of fear and fighting took over the Western world. Men banded together in small strongholds and defended their lives from other men. Carts and wagons were still used, but no one took his hand from the sword hilt long enough to draw pictures of them. For six hundred years most travel was on horseback; it was safer that way. Presently it came to be considered sissy for a man to ride in any kind of vehicle. It took a long time to change that idea.

The hammock-wagon appeared near the end of the Dark Ages. It was probably used for invalids only, but it must have been fairly common, because it appears in several of the early chronicles. The first drawings found showed the wagon only, and it was thought the thing might be some kind of bed on wheels, but later one turned up showing horses attached.

The hammock-wagon on the next page is shown starting down a hill. The horses had no way of holding it back, so it is shown being snubbed with chains to keep the rear wheels from turning. This was the only kind of brake then known, and an improved version of it was used almost as long as journeys were made behind horses.

An invalid could ride easier in a horse-litter, but there is no evidence that it was known in England before the Normans came over in 1066. It's a simple idea: a

Hindu gujerat (*cart*)

25

chair or bed on poles, with a horse to carry each end. Though there was considerable jouncing, there were no bumps. The Romans used them, and the Normans had probably had them for some time. They were in very general use in England from the Conquest until King John's time, about 1210. As late as 1640 a horse-litter was still the best way to move a badly wounded man.

The whirlicote opposite was for women of rank. It is illustrated in a psalm book made for Sir Geoffrey Lutrell about 1300. The horses were hitched one ahead of another, as they still were for heavy hauling in the seventeenth century and often are yet on English farms. Drivers rode the horses, acting as what were later called postilions.

Shafts or a pole must have been used on this wagon so that it could be steered, but nothing of the kind shows in the old picture so it has not been added here. Of all the horses in this book, so far, these had the best chance to use their strength and weight.

The wagon was slowed going downhill by lashing or chaining the wheels to prevent them from turning, just as with the hammock-wagon. You can see the chains hanging on a hook between the front and rear wheels. The three sticks projecting from the back of the wagon could be dropped to keep the wagon from rolling backward, so that the horses could be rested part way up a hill.

The use of carriages was increasing. In 1294 Philip the Fair prohibited Flemish citizens' wives from riding in them, with the idea of curbing luxury. Like most prohibitions it didn't work very well. At about this time a ballad about Queen Eleanor of England appeared. In it was a bit that said, "She was the first that did invent, In coaches brave to ride"; however, nobody saved a picture to prove it.

Many nations claim the invention of the coach. To judge by its name and its early date, the Italian entry, the *cochio,* might seem to be the original, but it wasn't much different from any other carriage-wagon of the time. However, the Italians have a better claim than this to precedence in coaches: at a time when there were but three in all England, sixty coaches took part in an Italian procession. The cochio was hooded with matting, a part of which could be rolled up for ventilation. The driver, perhaps from a dim recollection of Roman ways, rode on the wagon and drove with long reins.

Norman horse-litter,
about 1100

26

The remarkable thing is not the wagon but the harness of its horses. The traces are attached to hames, fitted on a padded neck collar, and are supported by a saddle and a breeching around the horses' buttocks, which helped them hold back on a downgrade. In a general way, this is the standard form of heavy harness now.

Though smaller, the Flemish carriage-wagon on page 28 is classed with the whirlicote and the cochio. New were the double-trees by which the horses were attached. This equalized the pull of the two horses and is still used, almost unchanged. It was always popular on the Continent, but the English didn't care much for it. The double-tree itself is a longish wooden billet swiveled on a pin which passes through the pole near its wagon end. Two shorter sticks called singletrees or whiffletrees are linked to its ends, and to each end of each single-tree a trace is attached. To permit sharper turns, the front wheels of this wagon were smaller than the rear ones, as they were later on all coaches and on most carriages. This required a transom, fastened to the bottom of the body over the front axletree, to keep the body level and to hold the kingbolt.

These Flemish horses, black and heavy, had wide wooden hames on their collars. There were pole straps from the bottom of the hames to the pole tip by which the

Italian cochio, 1288

horses could hold the wagon back, but there was no breeching such as the cochio horses had to help them to do it. For pulling there were rope traces which were simply passed through holes in the hames and knotted. These are the first horses we've seen wearing blinders to keep them from shying at objects along the roadside.

Travel on horseback still was easier on the human frame than riding in any wagon. When women rode horses they usually sat on *pillions*, which were little side-seats behind men's saddles. Instead of stirrups a pillion usually had a little strap-hung shelf

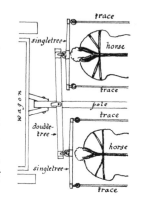

Modern neck collar and hames for work harness

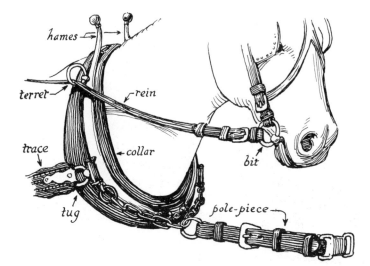

Flemish carriage-wagon, 1347

on which the rider's feet rested. Pillions were used occasionally as late as a hundred years ago. Before 1388 there seems to have been no other way for a woman to ride a horse, since for her to straddle one would have been scandalous. That year Anne, queen of Richard II of England, invented the sidesaddle. No doubt the argument as to whether it harmed the horse started right then; it's been going on ever since! A modern sidesaddle has only one stirrup, on the near side; the rider's left foot rests in this, and her right leg is supported, on the same side, by a padded wooden horn under her knee. The horn was an eighteenth-century addition; Queen Anne's saddle was simpler and more like a pillion.

Isabella of Bavaria rode in a throne litter for her formal entrance into Paris to join the French Dauphin whom she had married by proxy, sight unseen, and who slipped into the crowd in disguise to catch an advance glimpse of her. Her conveyance opposite looks imposing, but it was top-heavy and dignity ran a fearful risk riding on it. Probably Isabella had a safer enclosed litter for the road.

As for roads, there literally were none as we know them, just rough tracks over the countryside. They made no effort to get anywhere in a direct line, and their destinations were unmarked by any kind of sign. They ambled about, skirting fields and detouring to reach every hamlet along

Woman on pillion

Woman on sidesaddle

Throne horse-litter of Isabella of Bavaria, 1385

the way. Maps of a sort existed, but they were for libraries; it is doubtful that anybody ever thought of using one of them to find his way or that it would have been much good to him. The hired guides generally overestimated their knowledge of the way. People, nay, whole *armies* were often lost.

Travelers were fair game for thieves, so people going in the same direction tended to band together for protection and company, as did that notable band of pilgrims to Canterbury preserved by Chaucer in amber. Since each litter-passenger and each horseman of any consequence had a mounted servant or two and a couple of sumpter-mules to carry luggage, one of these traveling groups could grow into quite a caravan.

Isabella probably used a mule-litter for most of her journey from Bavaria. Mules rather than horses carried it, because mules are more sure-footed than horses and also because a mule has an easier natural gait. The body of such a litter was usually covered with leather and in bad weather could be completely closed, its occupant riding in total darkness but reasonably dry.

The French royal chariot belonged to the "Most Christian King" Louis XI. A Ro-

French mule-litter

French royal chariot, about 1465

29

Riding harness, about 1450

man would have called it a *carpentum* and, except for its fancy Gothic wheels, it was very like one. The King's dignity was far too sacred to allow him to drive a horse, though he might ride one without loss of face. There was no place on the cart for a coachman, so the driver added to the burden of the horse by riding as he drove.

The first experiments toward easing the torture of carriage-riding had already begun. In Paris in 1457 Louis's mother had a wagon which had its body suspended on chains, the first of the kind ever seen there. It was admiringly called the "wobbling chariot." As late as 1550 there were only three carriages in Paris. They were called *carrosses,* and you may guess that they looked a little like the carrosse on page 31, but you *could* be wrong. Two of these conveyances belonged to ladies of the royal court; the third seems to have been used by one René de Laval, who was too fat to mount a horse!

Saddle horses of these times wore very elaborate harness (now we call it "tack").

They sported quantities of unneeded leather and many tassels, fringes, and brass buttons. Saddles were high in front and very high in back, copying the knights' tournament saddles which reached halfway up a man's back to help him stay on his horse under stress. Looking at the drawing, it isn't hard to see whence the modern Mexican or Western saddle descended.

It was felt to be necessary in the fifteenth century to hold a saddle in place, not merely by a girth around the horse's belly but also by additional straps passed around him fore and aft, and usually by still another fastened to a crupper (crouper) held under the horse's tail. The crupper has persisted in modern carriage and wagon harness.

The first English-built coach was the work of Walter Rippon, who made one for the Earl of Rutland in 1555. Nine years later Rippon built for Queen Elizabeth I a coach which, in old prints, looked like the one on the next page, but it's hard to be sure. It was described as a "hollow-turning coach," which would mean that there was space for the front wheels to turn under the body, but the prints show nothing of the sort; no curtains show either, but it must have had them, because there's a description of the Queen having them opened, "that all her subjects present might behold her."

Sir Thomas Chamberlayne, who died when Elizabeth was young, is said to have brought from Flanders "the first coaches and the first watches that ever were seen in England." It is recorded that before his time vehicles under the names of chares, chariots, carroches, and whirlicotes were used in the island. The whirlicote has appeared in this book; the chariot we have just seen in a French example; two of Henry VIII's wives had them and so had

Queen Elizabeth's French coach, 1584

30

Queen Elizabeth's English coach, 1564

his daughters, Mary and Elizabeth. It is conceivable that the identical cart was used by all four. A four-wheeled chariot appeared later. It was entirely different. Chares are mentioned by Chaucer (about 1375); they seem to have had wheels, to have been horse-drawn, and to have been enclosed. Carroches must have been different from coaches, since several writers speak of "carroches *and* coaches," but what the difference was is your guess.

In 1584 the King of France presented Queen Elizabeth with "an exceeding marvelous, princely coche." Possibly from policy, possibly because the new coach held six passengers, she let her ladies use it and,

for herself, continued to use the old one. Two of the six in the French coach sat sideways with their knees projecting into what was known as the "boote," of which this is the earliest example.

The old Dark Ages prejudice against men riding on wheels began to break down. First royalty tried it, then great nobles like the Earl of Rutland, and, at last, the lesser gentry risked the indiscretion, for that's what it was considered when, in 1588, Sir Henry Sidney "entered Shrewsbury in hys waggon with hys trumpeter blowynge, very joyfull to beholde."

Lusty Henri IV of France, who was agin' the established order anyway, had no hesita-

Carrosse of Henri IV *of France,* 1610

English one-horse coach, 1616

tion in using his carriage though undoubtedly he would have paused if he had known he was going to be murdered in it. Henri's carrosse had no door, just a fringed leather curtain which hung across an opening in the low, solid rail. In spite of the earlier experiments, there was nothing (more than a cushion) to ease the shocks between the ground and the royal bottom.

Coaches became quite common in London at this time, that is, as many as twenty might have been counted on the streets in a week. In 1610 a stagecoach line was established between Edinburgh and Leith in Scotland, not a very long line, since one town can be seen from the other, but many years ahead of any other such service. Travel by any coach was a grueling experience; they upset, they broke down, and they were constantly stuck to the hubs in mud and male passengers were expected to get out and push! Post horses were stationed at ten-mile intervals on English main roads in 1617, but not for coaches, only for the use of those who went horseback.

The one-horse coach shown was used in England in 1616. Its body was a little more enclosed, it had a baggage rack on the back, and it was drawn by the strongest horse that ever pulled a vehicle!

Hackney coaches for hire were introduced in England in 1625 and are supposed to have been named for the section of London where they were based. In time the name was shortened to "hack" and applied to any public carriage.

Because the streets of old London were mere alleys, the coaches were built very narrow to be able to get through them. Life was tough for pedestrians, who were squeezed to the walls and splashed with mud. In the narrow lanes commoners were expected to "yield the wall" to the gentry. There were arguments about this. The same principle prompted the polite custom, still to be seen, of a gentleman walking on the outside when with a lady.

Before coaches there was much travel by water; fleets of rowboats carried passengers on the Thames. The new hackneys cut into the business of the watermen and their complaints were loud, long, and bitter.

About this time there occurs the first mention of wheels in America. Governor

London hackney coach, 1625

John Endecott of Massachusetts received a letter introducing "one Rich'd Everstead, a wheelwright . . . a very able man, though not without his imperfections." The appearance of the vehicles he built and those of Claydon who followed him is not known; doubtless they were simple farm carts, not without their imperfections.

The crush of coaches in the narrow ways of London became a wrangling turmoil; so, in 1634, Sir Saunders Duncombe tried sedan chairs to relieve it. Sedans took less space per passenger and strong men were plentiful and cheap. Strong they had to be, for the first sedans were heavy. It seems as much as two men could do merely to lift one, but they were actually carried, the chairmen wearing leather shoulder harness to ease the arm strain.

Coaches for hire came a little later to Paris than to London. They were called *fiacres*. The name is said to come from St. Fiacre (he was Irish), the patron of all French cabbies, including taxi drivers. The third horse of the early fiacre was needed to drag it through the quagmire streets. This vehicle had bootes to permit the seating of six passengers, and it flaunted an innovation in the form of a lantern swung above the driver's seat. This seat was attached to the front of the body. A few years later, when bodies were suspended, the front wheels of coaches were moved forward and the driver's seat was detached from the body to go with them.

Duncombe's sedan chair, 1634

Before and long after the introduction of the fiacre, the *chaise-à-porteurs* or sedan chair was popular in Paris. By putting wheels on one end of it and hitching a horse to the other, the French invented the *chaise* (also called chair, cheer, and shay) which, with some changes, was widely used in England and America. In the original French version the chaise was a one-passenger carriage. It was entered by a door in the front like a sedan chair and, to get in, it was necessary to climb over a shaft or duck under one. One reason for hitching the horse so far out was to allow room for this maneuver; another was to take advantage of the springiness of the long shafts.

Meanwhile the London "siddan" carriers (they were usually Irishmen) must have found the Duncombe version too heavy, for chairs began to be smaller and of lighter build. They hit a wave of popularity about this time which lasted some twenty-five years, until the public began to get qualms about using "freeborn Englishmen as beasts of burden." However, though Samuel Pepys expressed pity for a

Parisian three-horse fiacre, *about* 1650

London sedan chair, 1655

French chaise, 1664

poor man who had to come all the way across London in a chair, he had nothing to say about the poor devils who carried his friend that distance! It appears that sedan chairs had glass windows from the time of their introduction, a good many years before glass was generally used in coaches.

The "wobbling chariot" of 1457 seems not to have been much imitated, and the idea was no doubt considered brand-new when it reappeared in the seventeenth century. In order to suspend a coach body, it was necessary to create a rigid platform, and this was achieved by connecting the front and rear axles by a pole. A single one was called a perch; double ones came to be curved and were called cranes. The body was slung on heavy leather straps from four vertical posts mounted on the running-gear. Other straps were looped around the perch to keep the body from swinging too much,

though even a little must have been too much! No passenger did any lolling on the cushions; every yard of travel was a struggle to stay on the seat and to avoid as many bruises as possible. France built the first real roads, but she didn't get to it until this seventeenth century was nearly over.

Though England was slower to build roads, she was ahead in establishing stage-coaches. Six were running in 1662 and shortly, according to an old writer, "a multitude." A stagecoach was a public vehicle which went to its destination by "stages," using fresh horses for each stage. Posting-inns, which already existed for equestrians, stabled these horses and provided beds and food for travelers. The stagecoaches of this time still had bootes in which two passengers rode sideways. These are reported to have been very uncomfortable. One young man wrote home that he had been in the boote all the way to London and as a result had had to take to his bed on arrival. Bootes probably disappeared before 1700.

French coach, about 1640

English leather-covered coach, 1696

English coaches could be wider after the Great Fire of 1666 burnt the heart out of London, because the streets were widened. One of the many complaints lodged against coaches was their lavish use of leather and the resulting rise in its price. Here's a coach that *really* used it. In addition to being suspended on leather straps probably six or eight layers thick, the entire body was covered with leather. It had leather sun-curtains too, instead of windows, though glass in coaches was fairly common by this time; in fact, thirty years before this, Lady Peterborough had failed to remember the new glass windows in her coach and had put her head through one of them to greet a friend.

The need for springs was obvious to everybody, but nobody knew how to go about applying them. At least one English experiment showed promise in 1665 but apparently was not carried further. The date of their first successful use is obscure. They were tried on a wheeled sedan chair in Paris in 1670 but a hundred years later they were by no means universal.

The carriage below, essentially a phaeton, was probably called a *calèche* in its day.

The word was not applied until 1788, but the type was older and this is the gran'daddy of a long line of phaetons. It was one of the first open carriages with four wheels since Roman times and, though still without springs, it was nicely planned for showing off a handsome lady in a handsome gown. It had more than a touch of elegance, a new thing in vehicles, mostly due to the light construction of its wheels. To be so light they would seem likely to have been held together by a continuous, shrunk-on tire, though the earliest date which can with certainty be assigned to such a tire is 1760.

Ancient and medieval metal tires were made in sections and riveted to the wooden rims; even the first continuous bands achieved only a fair fit and had to be riveted to the fellies to keep them on. In that case the fellies had to be so made as to hold themselves together. Then some unhonored genius made an iron hoop which was actually *smaller* than the wheel it was to serve. To get it on, he expanded it with heat. It contracted as it cooled and gripped the wood so powerfully that little more was

French phaeton, or calèche de parc, about 1700

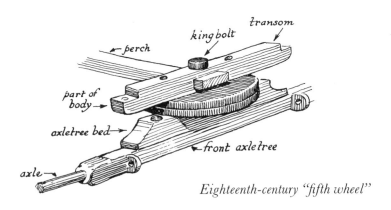

Eighteenth-century "fifth wheel"

needed to hold the wheel together. A light wheel built this way was as strong as a much heavier wheel of the pinned-together kind.

Our calèche-phaeton had a "fifth wheel," which seems to have been invented about this time. It is a device to steady the front end of a carriage while still permitting it to turn corners. In its eighteenth-century form it was two flat, metal disks lying one upon the other, centered on the kingbolt. The lower disk was fixed to the axletree bed above the front axletree, the upper to the under side of the transom which secured the forward end of the perch and was fastened to the bottom of the body. The bearing surfaces were greased and, in turning corners, the lower one rotated in contact with the upper. The principle is still used, but the "wheel" has been reduced to a skeletonized sector.

Sedan chairs gradually became lighter and more ornamental. In Europe there was never any break in their popularity. The French made constant use of them for nearly two hundred years. Their popularity revived in England with the dawn of "The Age of Reason" and they were used even in rural areas throughout the eighteenth century. They penetrated to the more elegant centers of the American Colonies where they were usually carried by Negro slaves.

Spanish sedan chair, about 1700
This is very similar to a French chair

French chaise, about 1710

English sedan cart, or chaise, about 1720

The first coach springs which had any general use were merely straight metal strips fastened by their lower ends to the vertical posts and having the leather braces hung from their upper ends. This, while a notable improvement, wasn't really very springy. When the new "S" or whip spring was invented (about 1700), it replaced the post entirely. Its lower end was bolted to the running gear and it swept upwards toward the body in a reverse curve. The braces were attached to the upper ends of the springs, just as they had been to the old posts.

The rig of the French chaise opposite was obviously better in design and finish than the original chaise, and it retained a major advantage of the first one: the bottom of the body was kept lower than the axles, which made for stability. In addition, the body was suspended by a fantastic cat's cradle from the new "S" springs. The means of entering the chaise seems about as awkward as concentrated effort could make it. The whole front dropped forward to rest on stanchions and hang over the horse's rump. The passenger climbed onto the shaft and stepped down into the carriage.

England adopted the chaise, but there it lost its elegance and became a plain, middle-class vehicle. As in France, the early ones were usually driven by a postilion. The small wheels of the English chaise made it necessary to bend the shafts upward so they would meet the horse at the proper level. Braking was accomplished, as in nearly all one-horse vehicles, by lashing the ends of the breeching to the shafts, allowing the horse to hold back with his buttocks.

The *berline* was named for Berlin, the city where it was first built. A form of coach, it had a suspended body which, in later examples, like the one here, was hung from springs. The berline had in effect two perches, which the French called branch-ards and the English called cranes. The great advantage was that these were deeply curved, allowing the body to hang close to the ground, reducing the tendency to upset. Coaches upset frequently, so the fairly stable berline became very popular as a traveling carriage on the Continent.

In the front of the book there's a drawing of a nobleman's coach-and-six which in the regular order of things belongs ahead of the 1755 stagecoach on the next page, the first of which there is a picture; though the minor gentry often rode inside, it carried far less pretentious people than did the coach-and-six.

Turnpikes had begun to be built in England. These were toll roads with no hard surface; they were simply graded dirt. On them more and more stagecoaches appeared; everybody began to travel. The prosperous rode inside; simple folk rode at the rear in the basket (it really *was* a basket); the daring clung, at low prices, to the roof with the luggage. Even though coaches often upset, due to the very high

French berline, about 1750

crown of the turnpikes, no one hesitated to pile weight on the roof, because it was universally believed that a high load was more easily pulled.

Beside the coachman sat a guard with a brass blunderbuss across his knees, for this was the day of the highwayman, and the black mask and wheel-lock pistol of Dick Turpin might appear on any lonely stretch.

The body of the coach was ornamented with "hip-raised" panels exactly like those used for the walls of rooms at that time. The sections of a coach body were ever afterwards called panels, even when they were perfectly plain surfaces. The terminals of the run were lettered on the lower panel of the door; the upper half of the door had leather curtains for bad weather. Outside passengers merely pulled their cloaks over their heads.

Ordinarily the whole contraption was pulled by two horses, with an extra, ridden by a boy, for muddy weather. Of course, the horses were changed frequently; the word *stage* once meant "stable." In descending steep hills, iron skids chained to the perch were put under the rear tires to keep the wheels from turning and to take the wear of sliding.

The gig was a one-horse cart which developed in England. The first gigs were badly designed, particularly from the horse's point of view; in addition to pulling, he carried just about an even half of the weight of the body and the driver. The gig shown at the top of page 39 had no springs. The back of the body hung from rigid iron stanchions; its front end was simply bolted to the thills (shafts). The folding leather top is the earliest one of which a picture has been found. It was known as a calash, and it had been invented in Italy some years before this.

Oliver Wendell Holmes wrote some verses about a "wonderful one-hoss shay" which he said was built of such superior materials that no one part outlasted another; at the end of precisely a hundred years the whole contraption disintegrated entirely. The date given above is the year in which it was supposed to have been built. The drawing at the bottom of the next page shows a shay (chaise), probably better known in its own time as a cheer—chair, that is.

It's likely that cheers were the first passenger carriages made in America. Ideas for them were taken from both chaise and gig, but the final results were as American

English stagecoach, 1755

as corn pone. Nothing remotely like the cheer's wooden springs had ever been made in Europe, nor was there any hickory there from which to make them.

The very early cheer was a little bit box-like in shape, a little rough in its workmanship, and its appointments were the simplest. Later models had folding tops, but the first cheers had standing tops of canvas laced to an iron frame. Seat cushions were canvas or cowhide (with the hair on) stuffed with straw. However, the weight of the passengers was over the axle and the whole rig balanced well and was easy on the horse. It's not surprising that this and the more finished later versions of the cheer were the most popular kind of vehicle in the country almost up to the time of the Civil War.

In the northern half of the Colonies everybody traveled on runners in the wintertime. The sleighs they used were simple, sturdy things built on the farm or in the shop of the local blacksmith. No doubt they varied somewhat between Massachusetts and Pennsylvania, but generally they must have been much like the one on page 40, from colonial Connecticut. To insure bringing home a full set of toes, travelers in such a sleigh rode with their feet on a couple of heated bricks or on a tin foot-warmer filled with hot coals.

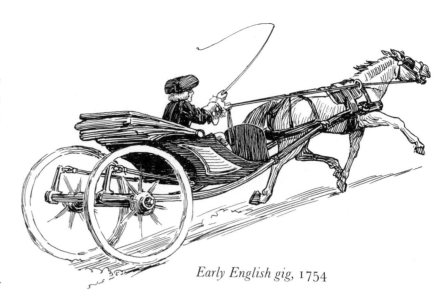

Early English gig, 1754

This Connecticut sleigh had but *one* shaft, on the off side. It's not clear why this was done or whether it was general practice. There's an old sleigh in the National Museum which has an ordinary pair of shafts set way over on the near side, so the horse could walk in the packed rut. That makes some sense, but the single shaft either left the horse in the middle, where he normally would be, or put him so far over to the right that he was off the road entirely. Since no one seems to have described the manner of hitching to a single shaft, the horse in the drawing has been placed where he could pull best, and his harness must be tagged "probable."

American cheer, or shay, 1755

American one-horse sleigh, about 1770

Colonial tin foot-warmer

Conestoga wagons hauled the freight of colonial America. The trails they made became the main highways of the eastern United States. In their heyday the Conestogas were pulled by six heavy horses. Five of the six wore iron hoops hung with bells above their hames. The sixth horse, the near-wheeler, had an ordinary saddle for the use of the waggoner, who drove from there or from a lazy-board projecting from under the wagon body and so placed that a man sitting on it could readily reach the long handle of the brake. The team proceeded at a walk and was controlled with a single jerk line. One long pull signaled a left turn, several short jerks meant turn right. In addition, the leaders obeyed the time-honored shouts of "gee" for right and "haw" for left.

There was no changing of horses; the same team pulled the wagon for the whole journey. They traveled by day and stopped at night at an "ordinary" which provided beds in dormitories and served plain food to wagonmen. These inns had large courtyards where wagons were parked. Frequently thirty or forty wagons a night put up at an important tavern.

American Conestoga wagon, about 1755

English stage wagon

The Conestoga wagon originated in Lancaster County, Pennsylvania. The legend that its dory-shaped body was the result of the first one having been built by a ship carpenter can't be substantiated. The shape was used because something like it was remembered from Europe and because it was a good shape to keep a load from shifting on a rough trail. The streams to be forded dictated the great height of the body from the ground. The hoops and canvas cover of the Conestoga were later adopted for the "prairie schooner," but the wagons themselves were different.

In England in the eighteenth and early nineteenth centuries stage wagons served approximately the same purpose that the Conestogas served here. They were even larger, riding on huge, coned wheels with tires nine inches wide, but the shape and construction were near enough to the Conestoga to establish a distant cousinship. They too had cloth covers over hoops but, instead of sagging, theirs were humped up in the middle. The stage wagons changed horses and kept moving night and day. They were pulled by eight horses, which were handled by a driver who rode alongside the team, mounted on a small cob or hackney. In addition to freight the stage wagons also carried passengers who couldn't afford coach fare.

They called the carriage below a barouche, but it would be a landau according to names later used. The nineteenth-century barouche had a folding head or top over the back seat only; this one had two heads as the landau has. This landau was perhaps the earliest formal open carriage, created by replacing the hard top of the coach with two calashes. These didn't fold so neatly and flat as later ones did but were inclined

English landau, about 1760

English brouette

The advantage of this was seen in England but, when they were introduced there (sans springs), the London chairmen screamed like panthers. A man was thrown out of work. Their compromise is familiar to our own time; one pulled and the other just went along for the walk! Perhaps this is why brouettes never came into really wide use in England, while as one-man carts they flourished forty years in France.

The French brouette on the next page had its axle-ends threaded, its wheel held in place by a nut and a washer instead of by the classic linchpin. Probably this was not the earliest vehicle with a threaded axle, but the idea seems to have started in France at about this time.

Some of the "Taxation Without Representation," which a little later cost King George most of his American colonies, no doubt went to build the gorgeous juggernaut below. It cost thousands of dollars in a day when money would really buy something.

The King's little wagon was made twenty-four feet long and was loaded with just about all the gilded ornament that could be crowded onto a coach that size, to a total weight of four tons! It really needed its eight cream-colored horses to get it over the ground. The English kings at this time were also the heads of the House of Hanover in Germany. There, these cream-colored horses were raised for their special use. When Napoleon invaded Hanover in the latter part of George's reign, he swiped a

to billows of excess leather. The tops of these early landaus were never more than halfway down, but they allowed grand ladies to take the air and show off their finery.

Elegant carriages had begun to run to decoration; almost any kind of picture might appear on a coach. This one, in addition to the crest on the door, had cupids with birds and flowers on the other two panels. The small panel below the door concealed a folding step.

The French put wheels under the sedan chair. When they hitched a horse to the result it was a chaise; when a man pulled it, it was called a *brouette,* sometimes a *vinaigrette.* The second name may once have been thought hilariously comic—a vinaigrette was ordinarily a box containing smelling salts. French brouettes had springs of various kinds and odd axle arrangements to permit the springs to work. Pulling a brouette was a lot easier than carrying one end of a chair, since the wheels supported the weight.

The state coach of King George III of England, 1761

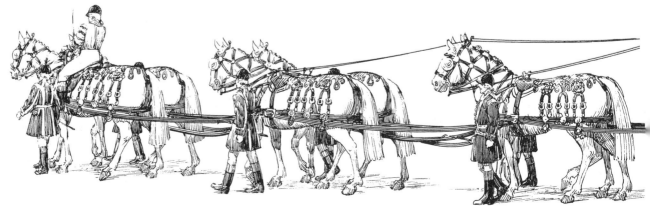

whole stable full of the horses and used them in his own coronation procession. George was so enraged that he switched to blacks.

The Royal Mews (stables in England are called mews) has several state coaches in it but this one is still *the* State Coach, used for coronations and for the opening of Parliament. It is known privately in the royal family as "Old Rattlebones." Springs and brakes have been added to it, and it was fitted with solid rubber tires for the coronation of Queen Elizabeth II, but its discomforts have not been eliminated, only eased a little.

The drawing shows the old shandrydan as it was originally built, with a high seat up front, covered with an elaborate hammercloth. Upon this sat a coachman who handled six of the eight horses; only the leaders were controlled by a postilion. Later generations have canceled the coachman, removed his seat entirely, and provided a postilion for each pair.

In addition to the four Beefeaters who guard the wheels, it is customary for three footmen to walk on each side of the coach. These have been left out of the illustration because they would hide the coach itself.

French brouette

In modern processions there is also a brakeman who walks behind. The harness for the coach horses was made of maroon leather with gold-plated mountings.

When George III's son was Prince of Wales he cut a dash as a sporting blood. One of his more spectacular diversions was driving himself and perhaps a friend to the races in a high-flyer phaeton. This was the hot rod of its time, an incredible vehicle, which like the Roman *carruca* had only one real purpose—to attract as much attention as possible. One reputable expert has claimed that this vehicle never existed and that contemporary drawings of it were caricatures, but the scathing descriptions written by people who saw it leave little doubt

English high-flyer phaeton, about 1770

that it actually looked about like the drawing. It took a fool's courage to ride in it, for it swayed violently and was obviously top-heavy, especially with the oversize Prince aboard. He used a ladder to get up and down. With all the high-flyer's absurdity it was a jumping-off point for a new era of lightness and elegance in coach-building which reached its peak some thirty-five or forty years later. In his obese old age the same gentleman had another phaeton built, for obvious reasons as low as possible. It is known as a George IV phaeton and was a direct ancestor of the graceful victoria.

In France they cut the front end off a berline body, leaving the gear as it was; the resulting carriage was a *berline-coupée,* often called a *diligence.* It was very popular with well-heeled traveling gentry. Usually it held two passengers, but some were built to carry one person only. The name diligence was also given to a clumsy stagecoach, quite different in every way. Very like the berline-coupée was an English vehicle known as a chariot or, if it had no driver's seat, as a post chaise. Later models of both kinds appear in this book.

Except for one experimental wagon which Cugnot had tried the year before, the wagon on the facing page is the first self-propelled vehicle, the ancestor of the automobile and the locomotive. It was intended to carry artillery. It never did, but it ran and still exists in a Paris museum.

The idea of a carriage that would move without horses had fascinated thinkers for a long time. Sails had obvious limitations but they were tried, and in 1475 a wind-mill-driven wagon was a failure. Leonardo da Vinci toyed with the idea of a spring motor and Sir Isaac Newton suggested a car driven by a jet of steam, but nobody built one. Low-pressure steam engines using pistons and cylinders had been invented by the middle of the eighteenth century and were in use for pumping water. It was time somebody tried one on a wagon. Cugnot did it and made it work—after a fashion.

His firebox and boiler were hung on a bracket, out ahead of everything, and delivered steam to two thirteen-inch cylinders. Each downstroke of a piston turned the single front wheel one quarter of a revolution, by the use of a ratchet. To steer, the

French berline-coupée (post chaise), about 1770

44

Nicholas Cugnot's self-moving wagon, France, 1770

engine and the boiler had to be moved with the front wheel.

The fire was stoked from the ground. The firebox couldn't be reached from the machine. There was no condenser. Steam which had passed through the engine was lost, so the boiler was soon emptied. Still, it worked! It hit two and a quarter miles an hour (with the wind behind it) and it could keep moving continuously for fifteen minutes. Its road career ended when it upset in a Paris street and a stampede was started by the noise and the escaping steam.

Cugnot's success didn't inspire anyone to abandon horses. Louis XIV had made a strong start at road-building in France but his successor did little to carry it on. Even the best French roads of this period would be considered almost impassable today, so traveling coaches were still made heavy and strong like the one drawn here.

This was the public diligence (stagecoach) which operated between Paris and Lyons;

it was the only diligence in all France at the time which was equipped with springs, and very unusual springs they were. Two of them, slung from straps, arched fore and aft, with the body balanced at the top of the arch. This put the door so far from the ground that a folding iron ladder was provided instead of a single step for getting in. The body had not only the usual tendency to sway from side to side, it also rocked forward and back like a hobbyhorse. An attempt was made to restrain both motions somewhat by four snubber-straps led from the running gear to the corners of the roof. The wickerwork at both ends was quite usual in eighteenth-century French vehicles. The entire bodies of some of the public stagecoaches were made of wicker.

A Roman pony cart was a *curriculus,* so the light English cart to which two horses were *yoked* was called a curricle. It is actually the only carriage since Roman days that had its pole supported by a yoke. The

French public diligence *(stagecoach),* 1771

English two-horse curricle, 1796

curricle was easy riding, well balanced, and well sprung, so it was deemed worth the five hundred dollars it cost. In the closing years of the eighteenth century curricles were ultrafashionable, and they remained popular for almost fifty years.

The earliest curricles, like the one in the illustration, had an unusually long pole with a rope stretched tight beneath it from the tip clear back to the axletree. The ends of the metal yoke or curricle-bar rested on the saddles of the two horses. A strap was looped over the middle of the bar and under the stretched rope; this allowed the rope to act as a spring to absorb the joggling caused by the motion of the horses. Later curricles used an actual spring. While we're speaking of springs—in an attempt to ease further the riding qualities of this and many other vehicles of the period, coil springs were often inserted as spreaders between the two parts of the leather braces on which the body was hung.

Behind the seat of the curricle, a sword case makes its first appearance in this book. At this time it was really used for carrying a sword, hence it was the mark of a gentleman's carriage. Because of this, the case re-

mained long after the sword went out of fashion, and the last ones, some fifty years after our curricle, were merely painted wooden blocks.

Here are also two more of the stylish carriages of this period, both with sword cases. The one on the left is a perch-high phaeton, a modified descendant of the high-flyer. It was intended to be driven for sport by its owner, carrying a couple of servants behind as supercargo. Such comparatively light carriages as this now stood some chance of survival on English roads. Attention was being paid to road maintenance, and experiments were being made to achieve a permanent hard surface.

The carriage on the right is a town chariot, tipped back on its rear springs in the latest manner. It is a refinement of an older form, and you can see how much it is like the French post-chaise diligence on page 44. Both the chariot and the perch-high phaeton clung to the traditional "S" springs. The emphasis in carriage design was now on elegance, which explains the appearance of the curved head-prop (for a collapsible top) on the chariot's immovable covering.

46

English perch-high phaeton and town chariot, 1796

For a body so far from the ground there had to be folding steps behind the chariot's door, and even so all but the most nimble might need the help of the lackey who rode standing between the hind wheels. The two horses of the chariot were in the hands of a coachman who sat upon his hammercloth as high as the top of the body and probably had to duck under low bridges.

Below his seat there is a black box, which also appears on the phaeton. This is the boot. Entirely different from the sixteenth-century boote, this one was a luggage-package-tool compartment. Some such box under that name persisted on fine carriages until the end of their time. The space that we call the trunk in an automobile is still known in England as the boot.

Importing English carriages was forbidden by the Congress during the Revolution. This gave a boost to American coachbuilding, and its quality continued to improve after importation was resumed at the war's end. George Washington had an American chariot built for his use in 1790; unfortunately, it has disappeared.

In 1798 the General imported a fine English coach. It met its end at the hands of Bishop William Meade of Virginia, who broke it up in 1802 and sold the pieces to finance his charities. Happily, Washington's friend, Mayor Samuel Powel of Philadelphia, had ordered a coach at the same time from the same maker, and his has survived. Except in the matters of ornament and paint-work the two were identical. David Clarke, who built them, came to this country with his carriages and set up business permanently in Philadelphia.

Under the able direction of Lt. Col. Paul H. Downing, the Powel coach has been painstakingly restored to its original condition and is now exhibited at Mt. Vernon. Of the many formal coaches used in eighteenth-century America, only the Powel and the beautiful Beekman coach in the New York Historical Society have survived. Colonel Downing has charge of rebuilding and equipping all the carriages at Williamsburg.

The Powel coach, at Mt. Vernon, 1798

American coachee, about 1795

Between the few who could afford such coaches and the many who rode in cheers, there was a sizable group of solid citizens who owned coachees. Also, families who had coaches often kept coachees as utility carriages. A Philadelphia tax list of 1794 mentions 33 coaches, 157 coachees, and 520 cheers; there were a few chariots and phaetons, too.

A coachee was lighter than a coach and of simpler construction. The drawing for this book was made from the only surviving example of its kind, now in the National Museum, at Washington, D.C. This carriage was once exhibited at Mt. Vernon but was discarded, foolishly, since there is reason to believe it was actually used there by the Washingtons.

Most coachees had side doors, but passengers entered this one through a door in the rear and sat facing each other on seats which ran lengthways. Instead of expensive glass windows there were leather curtains.

Coachees were used until about 1830; by that time, though they still kept the curtains, they were more pretentious, having a partition behind the driver, a sword case, and a platform at the back for a servant. The coachee was the first spring carriage to seat the driver on the body proper and the first to provide a shelter for him.

"C" springs, so named from their shape, were invented in 1780. They were more resilient than the old "S" springs. We've seen wooden cantilever springs on the cheer; here are wooden "C" springs. The leather braces on which the body hangs are carried clear over the outer surface of the coachee springs and, in order to make everything look as elegant as possible, the sides of the bent hickory are carved to imitate the leaves of a metal spring.

The steam road-engine below was not the first British attempt at a self-moving vehicle, nor was it the first attempt of its inventors, Richard Trevithick and a

Trevithick's steam road-engine,
England, 1803

cousin of his, Andrew Vivian, but it was the first one which can be nailed down as a practical success. A previous machine of the partners "sank into the road"; that's all we know, it just sank into the road! Another, which was framed of wood, burned up while the boys were in an inn having a snort of Christmas Eve cheer. It can't quite be proved that the third machine looked just like the drawing, but it hardly seems probable that anyone could have made up this model out of thin air.

Trevithick put a coach body on his engine so that the public would know it was intended to carry passengers, who must have needed both courage and agility to get into the thing! The boiler, firebox, and stack were at the rear. A shelf was provided for the engineer. This machine hit eight miles an hour on at least one trip of a little over a mile in London.

The steam cylinder lay horizontal under the body and the connecting rod turned largish pinions which meshed with gears on the drive wheels. These drive wheels were ten feet high! Steering was much handier than Cugnot's had been. The pivoted front truck was moved by a tiller long enough to provide some leverage. It must have been effective, because that first trip, from Leather Lane to Islington, was by no means a straightaway.

By 1805 Oliver Evans had built the first successful American self-moving vehicle, with a Greek name as weird as itself, *Orukter Amphibolos,* which is said to mean "Amphibious Digger." Basically the Digger was a steam-powered, paddle-wheel dredge

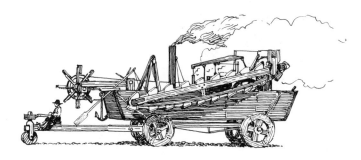

Evans's Orukter Amphibolos, *American,* 1805

ordered by the city fathers of Philadelphia. Wheels were put under it, and its engine was belted to one of them in order to get its twenty-ton bulk down to the Schuylkill River for launching. Evans must have had merits as a human being. When he ran out of money his helpers stuck with him and finished the job without pay. In order to reward them he exhibited the *Orukter* "moving around the Centre Square" for several days, accepting twenty-five-cent contributions from those who came to see it. There seems no reason to believe that the *Orukter* was not a practical dredge, once it had finally shed its wheels and gone overboard.

The wagon below was used as a country utility carriage. The "pleasure" may be hard to see. There were no springs whatever under the body. The removable seat, however, bounced at the ends of two hickory cantilever "springs" which rose from a wooden frame lying loose in the bottom of the wagon. The form of the body suggests a descent from the Conestoga wagon, though no connection between the two has been traced; nor can it be proved that the pleasure wagon is an ancestor of the buggy (page 59), but the eye seems to see a family resemblance. It's possible that the body

American pleasure wagon, about 1813

49

construction (with the framing outside) was copied on some of the later rockaways and, proceeding from there, may still be detected on the metal bodies of automotive station wagons.

Springs were used on the later pleasure wagons, but there's some question as to how much improvement they made. They were shaped like the springs of the French stage-coach diligence, on page 45, arched from front to rear with the body balanced on the top of the arch—but the pleasure wagon had no straps to snub the pitching!

Illustrated here are two gigs. Though one is American and the other English they are not shown primarily as national types but simply as two different styles of gig. The standard gig had wheels over five feet high, and they leaned outward from one another at quite an angle. It was well suited to rough roads. The American example is exhibited at the National Museum. It was built about 1792, but similar vehicles were in use until after 1830; in fact, this one was used long after that. It was considered a light cart in its day and, as with all gigs, only one horse was hitched to it, but it seems pretty heavy to a modern eye. Its iron "C" springs made it a far more comfortable conveyance than the cheer.

The other gig is an English stanhope. Its principal advantage lay in the low wheels which made it safe and handy to get into. Of course, the small wheels required a smooth road. In order to allow it to ride level the stanhope's shafts (thills, if you wish to call them that) were given a sweeping reverse curve which brought their front ends up high enough to be attached to the horse's harness. The body was set on what were called platform springs laid across the axletree. This was quite a departure in a day when nearly all carriage bodies were hung on straps. The lamps, set low on the sides of the stanhope, weren't only for swank. They were intended not to illuminate the road but, as with a ship's running lights, to show others where the stanhope was. The stanhope gig has always been a "smart" vehicle. Originally it was "correct" to hitch a piebald (black and white) or skewbald (brown and white) horse to a stanhope. This gig became popular in the United States, though in 1878 Stratton wrote that it had been "long since laid aside." It didn't stay aside, however; it revived, and its descendants, not too greatly altered, can still be seen at horse shows behind highly educated horses.

The widely used cabriolet was quite like a gig. It originated in France and was the height of fashion in England. It became popular over here, especially in the larger towns. Its wheels were higher than the stanhope's but not nearly so high as those of the standard gig, so its shafts had to curve upward a little to reach the horse. It had "C" springs from the first, and after

American standard gig, 1790–1830

English stanhope gig, about 1815

American cabriolet, about 1818

a few years the ride was further eased by setting the rear ends of the shafts on cradle springs fastened to the axletree. The illustration shows the earlier version.

For some years cabriolets were the vehicles most commonly offered for hire in towns, so some of them had a little hinged jump seat, just back of the dashboard, for a driver who sat almost on the knees of his patrons. The seat could be folded forward out of the way. In time the name "cabriolet" came to be applied to any one-horse hack and soon was worn down by natural erosion to "cab." Speaking of words—the horse in the illustration is having his lunch out of a feed bag, the name of which has survived in modern slang. Some people speak of "putting on the feed bag" as a more difficult way of saying "eat."

The first steerable ancestor of the bicycle was the Draisine invented about 1816 by Baron von Drais. The Hobby Horse, on the next page, was the English version, apparently but little changed. It was also known in England as the Dandy Horse and as the Pedestrian Curricle. The last tag was copied from the smartest turn-out of the time, as (it is barely conceivable) we might

in our day have called the machine a "Pedestrian Convertible."

The Hobby Horse was steered by a handle bar, and its shape was quite like a modern bicycle. Its wooden wheels had no tires and there were no pedals. Propulsion was by pushing on the ground with the feet. The exertion required was much greater than that for walking and the slight gain in speed could not be maintained very long. It might be said that the machine was tireless but the rider was not. Riding the Hobby Horse was considered sport. It had a considerable but short-lived vogue, and "riding academies" were established to instruct the young bloods in its navigation.

An Englishman named Louis Gompertz (yes, Englishman) attempted an improvement on the hobby in 1821 by rigging a ratchet on the front wheel. The rider operated it by leaning against a padded support and sawing a handle bar forward and back. It didn't become noticeably popular.

Except for the Roman roads, which were paved with cut stone blocks, and some of which are still in use, the first hard-surfaced roads in England weren't too successful. Not only were they crowned too high, but

Hobby Horse, or Pedestrian Curricle,
about 1819

equipped with no weapon more deadly than a key bugle, upon which he usually played expertly. Schedules came to be very rigid and speed was important. Changes of horses were made in a minute or so. The London-Edinburgh stagecoach *averaged* ten miles an hour for the four-hundred-mile trip.

The highest fare was still charged for seats inside the coach, and it was the correct thing for ladies to ride there. Outside passengers were no longer necessarily poor; in fact, the roof was much the more dashing place for a gentleman to ride. The old basket on the back had evolved into a large, wooden luggage compartment, and the seat on top of it now faced forward. Packages and other articles were hung all over a stagecoach and the roof was loaded to the limit with luggage and express, but there was a seat along its forward edge. Some privileged passenger sat with the driver, who was himself a personage of some consequence along his route. Often he executed small business transactions between one town and another or carried confidential messages. In Washington Irving's *The Sketch Book* there is a fine description of coaching and coachmen of this period.

their surfaces didn't stay smooth. They were built with mixed sizes of stones, and the smaller ones sank into the ground leaving the big ones on top for wheels to bump over. John McAdam solved the problem; he "macadamized" the whole roadbed with stones of uniform size. By 1825 his roads were being extensively built and traffic on them was heavy.

The dangers of the road were considerably less than they had been; coaches didn't upset so frequently and highway robbery had become a rare, rather than a usual, occurrence. So, instead of riding up front and carrying a blunderbuss, the stagecoach guard now rode behind and was

Most stagecoaches were pulled by four horses, and so this one would be normally.

English stagecoach and post chaise, about 1825

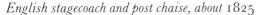

The extra pair in the lead, ridden by a postilion, were added because of bad road conditions or an extremely steep hill which will be met on this stage. The extra pair were called cock horses. Often a single cock horse was kept at the foot of a hill and was hitched on ahead of the regular team to help it up the grade.

There were no hand or foot brakes on these coaches. Their purpose was served by a skid, or drag-shoe, which was put under the near hind wheel going downhill. The guard handled the skid. A stop had to be made at the top of the hill to "skid up" and another at the bottom to "unskid." These skids were of iron and were attached to the perch by a chain. In use they became very hot, so, to avoid damage to the tire, it was supported by the sides of the skid, with a little air space under it. Some rural coach lines had no guards and the driver had to handle skidding. His reluctance to fool with it made him take chances which were the cause of many accidents.

If the stagecoach schedules were excellent, those of the mail coaches were magnificent; clocks were set by the horn of the mail coach. Before 1784 all mail was carried in England by mounted postriders who were often robbed or, worse, were themselves the accomplices of highwaymen. At best they tended to dally by the roadside. This casualness led people to write a futile *Haste post, haste!* on their letters and part of the phrase survives in common use. John Palmer persuaded the government to try forwarding mail by coach and the method was so successful that he was given the job

of establishing coach lines run specifically for carrying mail.

The mail coaches were built all alike, and all were painted with the royal colors: red undercarriage, with maroon and black body, having the royal arms and various heraldic devices on the panels. In contrast to the stagecoaches, which were painted in bright colors and which had their destinations and way stops marked all over them, the mail coaches had only *Royal Mail* and the two termini of their routes lettered neatly on the doors. Mail coaches carried a few passengers but no express packages. Mail coaches had numbers but no names. Nearly all stagecoaches had names like *Mercury, Tantivvy,* and *Tally-ho.*

The mail guard, wearing his scarlet coat and equipped with his three-foot tin horn, sat alone on the rear seat of the coach, his mail in a special compartment to which the only access was by a trap door beneath the guard's feet. In front of him was a box containing a blunderbuss, a cutlass, and a brace of pistols, but never in the history of mail coaches was it necessary to use them. The guard was responsible for the coach, the schedule, and the mail itself. He carried a locked chronometer and kept a detailed manifest. If an accident stopped the coach, it was the guard's duty to carry the mail forward by any means, in any weather.

Public post chaises were painted yellow. A post chaise with a "boy" (a small man) and a pair of horses could be hired for about a shilling a mile. This considerably more expensive than the stagecoach. Frequently travelers with the same destination, even though they were complete

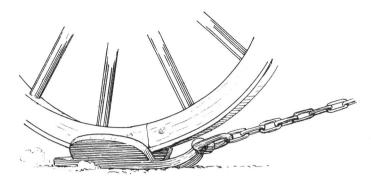

Skid, or drag-shoe

and nineteenth centuries to add class to coaching turn-outs. Mostly they ran behind the rear axle but sometimes behind the front one and sometimes, most spectacularly, under the pole-tip between the pounding forefeet of the wheelers and the flying heels of the leaders. It's recorded that one dog ran with a coach seventy-two miles a day on eight days out of nine! Occasionally he would ride a while to rest, but he preferred to run.

strangers, would share the cost of a post chaise. Post horses were changed at every stage and, in order to keep track of the vehicles, passengers, to their loudly expressed annoyance, had to shift to a new one at every other stage. The posting houses supplied horses for chaises and privately owned carriages only; coaches changed horses at their own stages.

Between the rear wheels of the coach a spotted coach dog is running. We call them Dalmatians, but their origin in Dalmatia is not proved and they have been called many names, including Talbot, Spotted Dick, and Fire-House Dog. It has always been their business to run with horses, though few get a chance to follow the trade nowadays.

Some think these dogs were in England before the Normans came, but it is known that many were imported in the eighteenth

Before the word "bus" was invented, the "Sociable" below was a public bus run by Mr. A. Brower on Broadway, New York. It was the second public carrier in local service in this country. Mr. Brower also ran the first one, which was just a large open carriage and was found to be too small for the job.

The "Sociable" had long side seats which probably held as many as ten passengers. Quite a staircase was needed to get into it, because the double-deck spring arrangement put the body so far from the ground. The "C" springs on which the body hung were set on a frame which itself rested on springs supported by the axletrees. The front one of this set was an elliptical spring, a form which in time brought about a radical change in carriage design.

American public "Sociable," 1829

General Lafayette's American calash, 1824

The route, from Wall Street to Bleecker, was not a long one, and the rather steep fare was one shilling for the whole trip or any part of it. Though fractional dollars had been minted since 1794, not enough were in circulation to provide all the small change that was needed, so shillings and pence remained in use for some years.

The boost that the Revolution gave to American carriage-building set it firmly on its feet and, though some people felt that their superiority demanded an imported carriage, what they rode in was not much if any better than the native product.

When General Lafayette revisited the United States in 1824 the government presented him with an American-made calash in which he made his triumphal tour of the States. The calash was named for its folding top; a little later such a vehicle was called a barouche. At this period ladies sometimes wore a collapsible bonnet which was also called a calash.

Lafayette's was an uncommon carriage even in its own time. In spite of the fact that it was somewhat cut under (that is, the front wheels could pass under the body in turning), it seems to have needed room to maneuver. On a narrow street in Marblehead, there's a house still standing which had one corner cut off to allow this carriage to turn.

The body seems almost absurdly small for such an equipage and equally small for its plump passenger. It probably rode well, however; notice the spiral snubbers which checked the bounce where the body hung on its leather braces. Notice also that the driver's seat, being a part of the body, had the "use" of the springs. When the top was up, it came well forward and gave good protection from the weather; when it was down, the great height of the body from the ground gave his enthusiastic fellow citizens an excellent view of the good general.

Along with the railroad experiments of the

James's locomotive coach and tilbury gig, 1830

Burstall and Hill locomotive coach, 1825

time, quite a few steam road-coaches were tried out in England to the loudly expressed horror of all lovers of horseflesh. All early steam vehicles were called locomotives, whether they were built for rail or road. Coal-burning steam lorries, still common in England, are ordinarily called locomotives. After 1830 a number of the locomotive coaches were making regular scheduled runs. The one on the previous page was built by W. H. James. It carried twelve passengers, and fifteen miles an hour was top speed. To an unsophisticated horse meeting this juggernaut on the turnpike, it must have been a fearsome thing indeed; one would hardly blame a modern automobile for shying at it.

The boiler and engine of the James coach were mounted on the rear, and power was applied to the hind wheels only. There was space for an engineer who also kept the fire going. There seem to have been no control levers for the steersman up front. It may be that he shouted over his shoulder to the en-

gineer for changes of speed. The only brake was the traditional drag-shoe, which was all right for skidding down hills but wasn't very helpful for a sudden stop. There was no differential gearing on the James machine, so getting around corners, with both wheels turning at the same speed, required skidding the inner one.

Another machine, built earlier by Burstall and Hill, did have a sort of differential in the form of spring-loaded ratchets which permitted either rear wheel to roll faster than its axle was turning. With this it was possible to coast around corners. This coach had also a reverse gear and, on the front wheel hubs, brake bands. It's possible that Burstall and Hill held patents on these features, so that James wasn't able to use them. The Burstall and Hill coach had a two-cylinder engine and was capable of four-wheel drive when necessary but, in spite of all its mechanical excellences, its best speed was only four miles an hour, so it was not considered a great success.

The little cart on page 55 to which the skittish horse is hitched is a tilbury gig, first built by a coachmaker named Tilbury, who, at about the same time, built the stanhope. It had the same low wheels, but its undercarriage rested on cradle springs like a cabriolet. The cutaway body was suspended at the back from a crosswise spring mounted on an iron stanchion, and from the front of the body two long "toe" springs were led out to the shafts. It was a

Gentleman's cabriolet, about 1830

very handy, comfortable little vehicle but heavier than it looks because of its seven springs.

The skill of the English professional coach drivers excited the admiration and imitation of sportsmen. Proficiency in driving became one of the correct accomplishments of a gentleman, as the spit-and-polish perfection of his rig, or turn-out, was the mark of his social position.

The man about town maintained a gig, a cabriolet, or a curricle, and several good horses; or he hired the whole outfit from a jobmaster. Wealthier men owned four or more carriages, each for a specific social occasion. With these went a stud, a stable full of horses of the various classes ordained to be driven with each kind of carriage. The horses were cared for by a small army of grooms under a head coachman "who might be mistaken for a duke, except that he dressed too well." Each rig demanded its correct harness, and a gentleman would be as likely to wear boots with his evening clothes as to use brass-mounted sporting harness with his town chariot.

The cabriolet originated in France. It was popularized in England about 1810 by Count d'Orsay and was improved by British coachmakers. As in America, there were common cabriolets and the type was popular as a hack, but, as the personal turn-out of a well-to-do bachelor, it was a thing apart and its every detail was dictated by fashion. For one thing, only a very large horse, over seventeen hands high, was allowed. A hand is four inches, which made the cabriolet horse five feet eight or ten inches at the withers, the ridge on the backbone at the root of the neck. In contrast, only a miniature groom five feet high, or shorter if possible, was permitted. Such a cabriolet groom was known as a "tiger." He rode standing on a board behind and,

to avoid knocking him off his shelf, the folding top was never put all the way down; it was carried "half struck," with only its front part folded. For really high style, a cabriolet was accompanied by two mounted outriders.

It cost money to turn out a cabriolet, but a smart curricle was even more expensive, most of its cost being in the matched pair of horses, which had to be exactly alike in size, in color, and in *gait*. With the pole supported across the horses' backs as it was, any inequality in their pace would joggle the passengers. This dictated the rope on the first curricle (see page 46) and the spring under the pole of the later ones.

The curricle was less flashy than the cabriolet, but it was elegant almost to the point of grandeur. Toward the end of its career, about 1850, the curricle often had a dickey seat behind for a servant but on the earlier ones he stayed home or stood up.

Business was so brisk on Broadway that in only two years Mr. Brower needed a larger carrier. It was called the *Omnibus;*

Curricle pole and bar

Curricle

the name originated in London. From it have descended generations of omnibuses and just plain buses. They multiplied quickly in New York; by 1835 the newspapers were railing at the wild maneuvers of the omnibus drivers.

Mr. Brower extended his route down to the Battery and up to Bond Street, reducing the tariff to twelve and a half cents. Fares were collected by a boy whose honesty was suspected from time to time. The *Omnibus* could seat about twenty passengers, and no doubt some stood up to create a tradition for their progeny. The elliptical springs which show in the drawing turned out to be too weak for all these passengers, and heavy, wagon-type, platform springs replaced them.

Approaching Broadway with its back to us is a chariotee, a member of the phaeton family, described as "a very aristocratic vehicle." It had elliptical springs set parallel to its axletrees. These had been invented in 1804 and, because they could be rigidly attached to both body and gear, they eventually allowed the perch to be abandoned on many vehicles; but at this date no one had thought that out.

The angle from which the chariotee is viewed shows the "dished" construction of its wheels. This was not peculiar to the chariotee; all wooden wheels were built this way to give them strength to resist side-thrust. Axles were bent downward just enough to compensate for the dishing and to use the maximum strength of any spoke in the bottom position by allowing it to stand vertical. English wheels were more radically dished than American ones, the difference being particularly marked after about 1830, when hickory began to be used for spokes in this country. This wood could stand more stress than oak was able to take, hence wheels could be flatter.

Just when the buggy appeared or how it came by its name is unknown. Even its parentage is obscure; it may derive from the pleasure wagon, or from the phaeton, or from both. In England a buggy was a gig with a folding top, and the American variety was described there as a "crude driving wagon." Even over here they were once known as top wagons and no-top wagons. The buggy shown here (which is certainly more phaeton than pleasure wagon) is a very early version of the carriage that came to be so universally used in the United States that, no matter what other conveyances a man might own, he usually had a buggy also.

Buggies were fairly comfortable; they were light and hence easy on horses; they

Early form of American buggy, 1827

were short-coupled so that they didn't need much space for turning, though (except for a later cut-under type) they could not turn sharp without cramping the front wheel against the body and upsetting. Buggies were strong, too; they are still made with perches. The first ones were entirely open; later they were more often made with folding tops, so often, in fact, that any folding top became a buggy top to an American. There was however a type with a fixed top, which later came to be known as a Jenny Lind buggy.

Fires happened even more often in early days than they do now, because open flame was used for both heat and light. Insurance groups were formed which paid "bucket brigades" of young men to protect the property of subscribers. If the burning house did not show the iron marker issued by its sponsoring company, the brigade just stood around and watched the fun.

From about 1800 volunteer fire brigades were formed who would fight anybody's fire. These became social clubs, with intense loyalty to the brigade and intense rivalry between brigades. A house could

burn down while the boys slugged out the question of who should turn a hose on it!

The volunteers used pumping engines like the one shown. When the fire bell rang the members grabbed their hard, varnished hats and the towropes and dragged the machine to the fire as fast as they could pelt. They put a suction hose into a pond or a well, and most of the members applied themselves to the pump handles while a smaller group dealt with the fire.

The handles were carried hooked against the walking beam of the pump. In action they were held in clamps and projected on either side of the engine. The pumping was done on two levels, part of the crew on the ground and the rest standing on the springboards which were extended for the purpose. With this system some twenty-four men could work at the same time.

There was no refrigeration in those days; nobody paid much attention to sanitation; flies were nothing worse than a nuisance; so butchers delivered meat in open, shallow-bodied carts. Whether is was the need to get the meat to the customer quickly, before it spoiled, or was only natural deviltry, butcher boys were proverbially wild drivers.

Grocers had used carts, too, but at about this time they began making deliveries in a light, open, spring wagon, of the kind which later became known as an express wagon. The springs were elliptical. This design, though no longer used by grocers,

American hand-pumped fire engine, about 1830

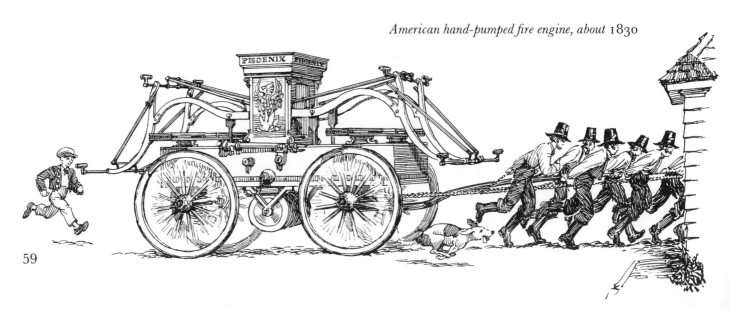

American grocer's wagon and butcher's cart, about 1838

has never changed much and may still be seen in use by vegetable hawkers or junk collectors.

From the eighteenth century onward it was customary for wealthy young men to finish their educations by taking the grand tour of Europe. Older people, of course, traveled there also, on business or pleasure. Such was the ruggedness of some parts of the journey that special carriages came to be built for it. From 1770 on it was fashionable in France to copy English coachwork, though at that time the French were still doing the better work.

These carriages were of several types, of which the *dormeuse,* or closed britzska, was perhaps the most elaborate. It acquired its name from the fact that the seat cushions could be arranged to allow a passenger to sleep at full length, his feet extending forward into the boot. Under one cushion a commode was concealed. The interior of the dormeuse was lighted by candle lamps, there was often some means of heating it, and equipment was carried for cooking food

along the road. A king's messenger sent on a diplomatic errand, his pouch under the floor boards destined, say, for Russia where the going was rough, might be glad he had a dormeuse rather than a common traveling coach.

The cooking gear, horse feed, tools, and such road needs were carried in the boot. The owner's personal belongings rode in a leather "imperial" which covered the whole roof. The sword case or something like it, which also carried pistols, was accessible from the interior of the carriage. Servants rode in a hooded dickey seat behind and kept warm as best they could.

Four horses under postilions were needed to pull the equipage, and a professional courier preceded the whole outfit by about twelve hours to arrange for inn accommodations and changes of horses. There were no brakes on a dormeuse; instead, it used not one but *two* drag-shoes plus a safety chain which hooked around a wheel spoke. There was also a drag staff on the rear axletree to keep the carriage from careering

French dormeuse *traveling carriage, about* 1840

English mail phaeton and cocking cart, about 1840

backwards to the bottom of a steep Alp.

Back in England the sporting squires developed various vehicles adapted to their pursuits and incidentally a bit on the showy side. The admiration for coaching influenced their design. Two of these, the mail phaeton and the cocking cart had "mail axles," copied from those of the mail coaches, and platform springs, which were then called "telegraph springs," from the name of the first stagecoach that used them.

The mail phaeton was named for its axles, which were designed to retain grease. Their chief visible characteristic was a flat hubcap held by three bolts which passed all the way through the hub. This vehicle was the station wagon of its time. Pulled by a pair, or by four, or by a spike or unicorn team (three, hitched two and one), it met guests arriving by stagecoach and had space for their luggage; or it carried fishermen or hunters and their equipment.

The cocking cart was more specialized and less reasonable. Apart from swank, its main purpose was to transport fighting chickens in the high box which was its body and on top of which its human passengers rode. In effect the cocking cart was a two-wheeled coach with everything omitted except the driver's seat and the guard's. It was invariably driven tandem, that is, with two horses hitched in single file (see page 86).

The clarence was named for an English duke and was used on both sides of the water as a private carriage and as a hack for hire. There was no difference between the two kinds except their condition; private vehicles ended their lives as "flies." The two seats facing one another inside the body could accommodate four passengers. Later models often had a curved glass front like a show window. The clarence in the illustration below had a concealed step.

This vehicle was one of the first carriages to be built without a perch, the front and rear axletrees being connected only by the body itself, bolted to four elliptical springs set fore-and-aft. The front wheels were cut under, that is, the body was made high enough in front to allow them to pass under it. As a result, the carriage could literally be turned around in its own length.

In addition to cabriolets, which could still be hired on American streets, a rear-entrance cab with side seats appeared. It was quite like a London cab of the time

American clarence, about 1840

61

American rear-entrance cab, about 1835

American sulky, about 1835

called a Boulnois. Though the local version was never very popular, it could carry four passengers and a driver who sat on the roof. The old pictures are not clear. The axletree may have passed right through the body, it may have been bent, "cranked," to pass under the body, or it may have been a mere stump of an axletree, sprung to the body itself. Cabs have been built all three ways.

The English of the early nineteenth century tended to call almost any one-man vehicle a "sulky," even applying the term to a four-wheeled, single-seated chariot of the kind known in France as a *désobligeance.* An English doctor, one of the prolific Darwins, is supposed to have invented the sulky road-cart to save time by making it impossible for his friends to ride with him on his rounds. He simply had a narrow seat put on a whisky cart. The American sulky shown here is much like the Doctor's whisky except that its wheels are higher and it isn't as nicely sprung. Sulkies were light and easy on a horse, which recommended them to contractors, horse traders, and others whose work required getting over a lot of country.

A word about dates—few individual carriages or wagons were important enough for any record to be kept of them except by the builders, and few of their account books and design books have survived. Types developed and were used over long periods, changing slowly and but little. The brougham, for instance, was in use for at least three quarters of a century. Usually the only way to date is from the earliest picture that can be found. Since the pictures tend to occur in bunches, gathered into a single book, whole groups of vehicles have the same approximate date.

English steam phaeton, 1838

Walter Hancock was a leading builder of steam coaches, which now began to run quite regularly on English roads. Beginning in 1828, he built half a dozen big ones, each of which had a name. One which ran from London to Stratford (100 miles) was startlingly called the *Autopsy!* In 1835 Hancock built the *Automaton,* which seated fifteen people and went up a steep hill at seven and a half miles an hour. The little phaeton on the opposite page was Hancock's own car. Probably it is the earliest of all private automobiles. On it as on all of Hancock's coaches, the engine was enclosed in the body.

At first England was miles ahead of the world in the development of self-propelled vehicles, but they soon collided with a stone wall of prejudice and self-interest. The operators of horse-drawn coaches, the horse-breeders, the "horsy set," the new railroads, and the people who just wanted no part of anything new were all out to get the steamers and get them they did.

At first there was just sniping and then attempts to induce "road accidents," but presently organized lobbying forced through stringent laws which controlled steam coaches right out of existence. For instance, it was legal for a steam vehicle to use a public road *if* preceded by a man on horseback carrying a red flag! The inventors and promoters quit, and England faded from the automotive picture until the laws were repealed in 1898.

Lord Brougham felt the need of a one-horse closed carriage in which to get about London. After failing to persuade his regular coachmaker to build anything so radical, he found a man who made him an improved copy of a French coupé. It proved to be very convenient, "a carriage for a man who carried his own carpetbag on occasions." There were seats for only two passengers, and the driver's seat was attached to the front of the body instead of riding ahead over the front wheels. The brougham served the same purpose as the heavy old chariot, but it was handier and it saved the cost of one horse.

Because it was practical and because His Lordship's use of it made it acceptable, the brougham was widely copied. It became the correct equipage for a wealthy bachelor in town. It also made a very good hack and, as such, came in time to be known as a "growler." Though in essentials they remained unchanged, there were slight differences among broughams in later years. Not only the makers' ideas but also conditions of use brought about slight modifications. A country brougham had slightly larger front wheels for dirt roads, a station brougham had a rail to hold luggage on its roof.

The vehicle from which the illustration was drawn is in the Science Museum in London, where it is exhibited as Lord Brougham's original carriage. Painted olive green, it has a sword case behind the seat.

English brougham, 1837

Trotting races have always been more popular in America than in any other country. In the first one recorded, in 1818, a horse named Boston Blue covered a mile on the Jamaica Road in three minutes. He was ridden by a jockey, as running horses still are. It wasn't until 1845 that a trotter first pulled his driver in a wagon instead of carrying him. By the way, though all horses who race without galloping are called trotters, there are some whose gait is not actually a trot. These are pacers, like the one pictured below. A pacer moves his front and back legs on the same side in unison, like an elephant; a true trotter moves his fore and hind legs on opposite sides together.

It's hard to say why a four-wheeled wagon was chosen for racing rather than a handier two-wheeled cart. The wagons were very short-coupled and, to keep them as light as possible, they were no stronger than barely strong enough. There was just enough frame to unite the wheels and support a seat. The spokes were about as thick as pipestems and a collision on the track reduced both wagons to a mess of splinters. Carts were used for most harness races after 1870 (see page 76), but wagons were raced occasionally up to 1902.

To indicate speed, our fathers spoke of "going like sixty"; our grandfathers mentioned a "two-minute clip"; but our great-grandfathers couldn't do better than "going two-forty." A horse called Uhlan pulled a wagon for a record mile in two minutes flat (that's two minutes exactly, neither plus or minus any seconds). It was a pacer, Dan Patch, who finally broke two minutes—within the past fifty years. The present trotting record for a mile is 1:55¼ (with a modern sulky, of course).

If the racing wagon was the lightest of mid-nineteenth-century pleasure vehicles, the lightest working vehicle was the covered business wagon. The first one was built in New York in 1844 by Ezra M. Stratton and was used to deliver bottled soda water. The protection it gave to merchandise made this style very popular. In the early years of the twentieth century city streets swarmed with such wagons, and the design hadn't changed very much from Stratton's original. It was, after all, only a flat, spring wagon with a fixed hood, and there wasn't much change that could be made in it.

Stratton has an interest of his own and must be given the credit that is due him. He was a successful New York carriage-maker and he became the editor of the *Coachmakers' Magazine*. In 1878 he gathered what he had learned at both trades into a book called *The World on Wheels*. Stratton was no mean scholar. His information on vehicles of his time and earlier is dependable and has formed an important source for the preparation of the book you are reading.

American skeleton racing wagon, about 1845

Stratton's "Improved Business Wagon," 1844

The carriage below, with six-foot wheels, was first designed in Spain, so it isn't remarkable that it was widely used in Louisiana and in Cuba and Mexico. The one in the picture was owned by Jenny Lind, "The Swedish Nightingale," when she was singing in Havana. The volante was always a lady's vehicle; a gentleman who wasn't crippled was expected to bestride a horse. Naturally no dainty and cloistered Latin lady could soil her hands by driving a horse herself, nor could she with propriety sit by the side of a coachman in so intimate an equipage. So the lady sat alone with her hands formally folded in her lap and the driver rode the horse, which otherwise didn't carry too much weight because of the great length of the shafts.

From about 1830 through 1870 volantes were built in New York in great numbers and shipped south. After that the trade was gradually lost to Paris and London.

The buggy's great popularity naturally caused it to be simplified and cheapened. Where the form shown earlier in this book had small front wheels and a certain sweep to the body, the later ones developed a flat box body and wheels of nearly equal size. Though there was also a "coal box" buggy, the square or box buggy like the one on the next page became the standard and has never changed very much except to become even simpler. Many a one of them is still on the road, and it is still possible to buy a spanking new one fresh from the hands of the builders, but not for the $22.95 that seemed to our ancestors to be the right price for a buggy.

No vehicle was more entirely American than the buggy. It had to the greatest degree the light strength, "the eccentricities of lightness, as Englishmen consider them, such as distinguish certain American carriages." So the Duke of Beaufort put it. The reason for this lightness is probably nothing more complicated than the fact

Spanish volante, about 1845

American box buggy, about 1855

that America had hickory wood and Europe didn't. No tougher, more resilient wood exists, and its use allowed American spokes, fellies, and frames to be thinner than the oak ones favored in Europe. For a long time fashionable Americans thought the heavier European work was smarter and did their best to persuade Yankee coachmakers to imitate it. On the other hand, there was at least one English coachmaker who felt that the British could learn much from American vehicles—especially from the buggy.

Abe Lincoln used the carriage below. He must have had real trouble with his long legs in its shallow body, especially when it was loaded to its full capacity of four statesmen. When only two were riding, the back of the front seat folded down to form a deck over the front part of the body. The occupants of the front seat faced backward. It's

a barouche because the rear seat has a folding top. Ten years later this would have been called a calèche, because that became the fashionable name for a barouche in this country. We lapsed to the old name again later. Barouche or calèche, it was probably one of the last carriages to have its body suspended on leather braces from "C" springs. It had elliptical springs too; on the roads and streets of its time it needed all the springs it could carry. The peculiar podlike object above the front wheels and under the driver's seat is the boot, where a few tools and a sandwich could be stowed.

There's a natural confusion in many people's minds between the older Conestoga wagon, which carried freight in the East, and its descendant, the "prairie schooner," which went with the new settlers to the far West. They were both "covered wagons" and at a glance they looked somewhat

Barouche, 1865

Prairie schooner, about 1860

alike, but the schooner had a flat, box body instead of the tipped-up boat-shaped affair of the Conestoga. The wooden sides weren't so tall in the later wagon.

Year after year, an unending train of such wagons crept westward, creaking forward at an ox-pace by day and warily camping at night. A family's whole possessions were in the wagon. Only the weak and infirm might ride, the able-bodied walked. This isn't the place to tell the story of the hardships and heroisms of that migration which helped to make America great, but here is the wagon which left its ruts on the plains, where they are still to be seen.

The settlers, once they reached "home," had to make do with what they brought with them or what they found on the land. One thing in short supply was iron. Craftsmanship was available, however, and it created the Red River cart, which was built of hand-hewn wood using *no metal whatever.* Its hubs and even its wheel fellies were lashed together with rawhide somewhat as Egyptian chariot wheels were lashed. This was no primitive vehicle; the men who built it knew all about carts. A buffalo hide was ordinarily carried. It could be used to cover the load or it could be lashed around the underside of the body to convert the cart into a boat for crossing rivers!

The velocipede, which was simply a Hobby Horse with front-wheel pedals, has been almost forgotten, but it had ten years or so of considerable popularity. The earlier

Red River cart, about 1860

ones were made almost entirely of wood, but later they had metal frames and wooden wheels. Their well-earned nickname, "boneshaker," has been incorrectly transferred by some to the "ordinary," the high-wheeled bicycle which followed the velocipede.

Pierre Lallement had worked in Paris for a Monsieur Micheaux who was interested in velocipedes, and it's just possible that the mechanic may have swiped from the boss the idea that he brought to the United States and patented. Lallement's machine was made largely of wood and had no brake whatever. His pedals could rotate on their pins, and they were counterweighted to help in keeping the tread uppermost. There was one other advance over the Hobby Horse: the saddle rode on a curved spring.

Lallement velocipede, about 1862

Van Anden's Dexter velocipede, 1869

Roper steam velocipede, about 1868

A man named Hanlon patented a velocipede of which no example survives, but he must be credited with being the first to suggest the use of a "rubber band" instead of iron for tires. Van Anden's Dexter (named for a famous horse) was based on Hanlon's design but didn't use the rubber bands. Its seat was far too low, giving the rider less mechanical advantage than he had on Lallement's earlier machine. The Dexter did have a brake, however, which pressed against the rear rim and was operated by twisting the handle bar. It also had a feature so far ahead of its time that it wasn't generally used until some forty years later: it was freewheeling. The front hub was ratcheted; the wheel could coast while the spool pedals remained stationary.

In Massachusetts S. H. Roper built and actually operated a steam-driven velocipede, and in Paris at about the same time, the late sixties, Lallement's old boss, Micheaux, did the same thing. The engine of Micheaux's machine was mounted above the rear wheel, which it drove with a belt. Roper's device had two cylinders, one on each side of the rear fork. They turned cranks on the ends of the rear axle. The boiler was somewhat uncomfortably located under the seat, and it was necessary to keep moving to outrun one's own smoke. There were no pedals, merely footrests. The brake was a friction affair similar to Van Anden's. Roper's was not a practical road vehicle because of the absurdly slight clearance between the firebox and the ground.

Large Western Concord coach, about 1870

Dudgeon steam vehicle, 1868

Richard Dudgeon built a ten-passenger steam vehicle in New York in 1868. But for its unflanged, steerable wheels, it was a locomotive. It is said he had built another ten years before which was destroyed when the New York Crystal Palace burned. The second one has not only survived, it's still in running condition! Power was delivered to the rear wheels by two separate one-cylinder engines, directly connected. The passengers kept their feet warm by resting them on the boiler. The "engineer," steering from a low seat in the rear, could see almost nothing of what was ahead of him.

Only a while ago, say when grandfather was a boy, there was a limited choice of places you could reach by railroad but, with enough fortitude, you could get almost anywhere by stagecoach. Nearly all of them were built by Abbott and Downing of Concord, New Hampshire, and hence were called Concord coaches. The first ones were made before 1830.

In the East, light Concords rode the turn-pikes. The heavier model followed the covered wagons westward in America and followed the northward trek in South Africa. No horse opera is complete without one and that's the way it should be, though they were originally introduced to reach the mining areas of California and Nevada.

These big Western Concords carried nine passengers inside on three seats provided with safety straps for rough going; and though roads of a sort were built, the going was rough. On a famous occasion Horace Greeley was given in full measure the speed he demanded of a Western stage driver and arrived at his destination incapable of appearing for a scheduled lecture. There were seats for six passengers on the coach roof and many more might cling where they could find space. A total of thirty-five souls riding on one coach has been recorded. The two seats alongside the driver usually had to be occupied by guards. If you didn't hanker for a *long* life, holding up gold-bearing coaches was the easiest way to live in the Old West.

The Concord coach was springless. The job of easing the ride was left to the two thoroughbraces on which the body was suspended. These ran fore-and-aft and were made up of eight or more thicknesses of oxhide. The iron stanchions which supported them were completely rigid.

The Western stages were almost always pulled by six horses which, rather surprisingly, seem usually to have been matched in color. Each pair was hitched to a set of doubletrees. Those for the wheelers lay

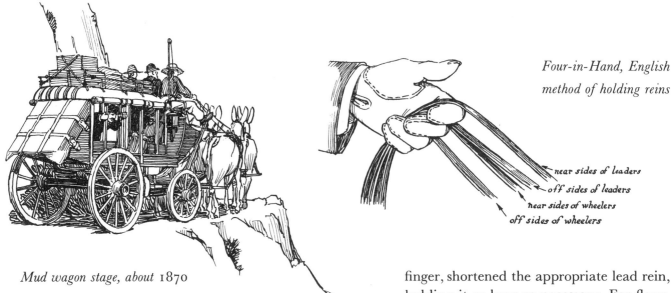

near sides of leaders
off sides of leaders
near sides of wheelers
off sides of wheelers

Mud wagon stage, about 1870

across the root of the pole, those for the swing horses were pivoted athwart a second pole hooked to the end of the first and carrying at *its* end the lead bars to which the traces of the leaders were attached.

The wheelers rarely had breechings. Slowing and stopping were left to the long-handled brake which the driver operated with his right foot. On long downgrades the wheels were sometimes chained.

The "jehu" ("for he driveth furiously": II Kings 9:20) managed his team from his seat with six reins held "American style," three in each hand, the near reins in the left hand, the off reins in the right. For a slight change of direction the reins on the desired side were simply pulled, all three together. This was called "chopping." To round a sharp corner, the driver brought his hands together and, with thumb and finger, shortened the appropriate lead rein, holding it as long as necessary. For flourish he had a whip with a five-foot handle and a lash not long enough to reach the leaders; a horse which needed to be whipped didn't belong in the team. These drivers were famous for drinking whisky and for their effortless control of their animals. Such prodigies are reported as turning a galloping team clear around in a street; obviously it was a wide street. Their modern successors who drive Concords for the movies don't do badly either at driving.

Stage transportation throve in the West from the early days of the Gold Rush until the coming of the railroad; it dwindled gradually to serving remote areas and finally disappeared completely in 1910. The tales of its dashing drivers, of dastardly bandits, of wild rides and freak accidents make entertaining reading. It is to be remembered that at one time travelers could take a coach at the end of the rails, at St. Louis, and travel *day and night*, changing drivers and horses, to San Francisco in 23 days.

To reach destinations beyond the merely remote, passengers transferred from coach to mud wagon, which was also a coach of sorts. It too was suspended on thoroughbraces. It was lighter than the Concord and cheaper, being built with simple flat sides. Like the coaches it had a covered boot at

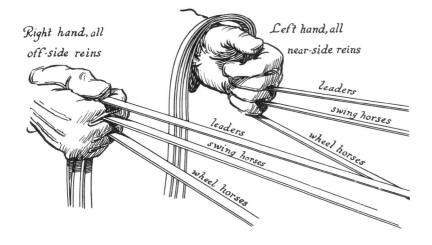

Right hand, all off-side reins

Left hand, all near-side reins

leaders
swing horses
wheel horses
leaders
swing horses
wheel horses

Arrangement of six-horse reins for "American style" driving

70

Forder's improved hansom cab, 1875

the back, but its body wasn't enclosed, except by roll-down canvas curtains. Often it was drawn by mules, as sometimes the Concords were also. There's not much about the mud wagon to suggest that it was any easier to pull through a slough than a coach; it probably acquired its name from the general condition of the roads on which it was used.

The hansom wasn't the first one-horse cab; and the hansom as originally designed appears to have had no public use at all. In the early nineteenth century ordinary cabriolets with drivers were offered for hire in London. The presence of the driver on the single seat with the passengers produced an undesirable intimacy and crowded the seat. A folding seat for the driver which sat him practically in the laps of his passengers proved to be little better. In 1823 the hackney cabriolet, later called a "coffin cab," was introduced. Like the fashionable cabriolet of the time it had a "half-struck" top and a curved wooden apron which covered the knees of its passengers; the driver was relegated to a precarious little seat over the off wheel. When this proved too dangerous, the body was narrowed and the driver's seat was put between it and the wheel.

At about this same time (1830) the Boulnois rear-entrance cab appeared in London. It was enclosed. Its driver sat on the front edge of the roof, and in general appearance it was so nearly like the American rear-entrance cab on page 62 that it's not necessary to illustrate it here.

Mr. Hansom was an English architect. He took the roof seat from the Boulnois and the wooden apron from the hackney cab and added an idea, which seems to have originated with a Mr. Moore, in the form of wheels seven and a half feet high. This startling vehicle he actually built in Leicestershire in 1834 and personally drove it to London. He obtained a patent on it and sold the rights to a coachmaker for 10,000 pounds, very little of which he ever collected.

Joseph Hansom's theories were good. The enormous wheels would ride easily on cobbled streets; the body was low, which made the cab easy to enter; and the horse bore little weight because the passengers, sitting behind and below the axle, tended to balance the driver. John Chapman, who bought the rights, built his hansoms without the oversize wheels. It was a man named Forder who finally assembled the characteristics generally associated with hansom cabs. His vehicle is illustrated. He seated the passengers above the axle and put the driver up behind where his weight tended (but failed) to lift the horse off his feet.

71

English hackney cabriolet, 1823

Hansoms became the most prevalent hiring hacks in London, but two attempts to introduce them in New York in the late seventies were complete failures. This astonishes people who can remember hansoms swarming on Manhattan streets. Just why the town would first reject what it later embraced isn't clear. The New York cabs were rather higher and squarer than Forder's. A few of them are still maintained as curiosities to be hired by tourists. One of them appears in this book on page 87.

The world pays scant attention to the sometimes great contributions of the fellow who came in second. Once George B. Selden was credited with being the inventor of the "gasolene" automobile; lately he has been largely forgotten. Neither the credit nor the neglect are deserved. Selden didn't experi-

ment with an operating vehicle; his claim was based on the model shown opposite, which was submitted to the Patent Office, but a full-size version built according to his specifications in 1906 did actually run.

Selden mounted his engine (which he didn't invent) to drive the front wheels. The body was cut under so that the whole power truck could be turned 180 degrees and, with the wheels still turning "forward," could be used for backing up! Selden provided for a clutch, a foot brake, and a muffler, all of which were ahead of their time; but Jean Lenoir had successfully operated a gas-wagon in France in 1863, years ahead of Selden's first model, and Siegfried Marcus in Vienna was ahead of him also. Quite possibly no one in this country had heard of those two when Selden was granted a patent in 1895 on "a road vehicle powered by an internal-combustion engine."

This patent was claimed to cover all gasoline-powered vehicles, and nearly all early U.S. manufacturers were licensed by Selden and paid him royalties for the privilege of building cars. Only Henry Ford refused to pay, so he was sued by Selden and the licensees in 1903. After years of talk a court decided that Ford didn't infringe the patent, since it specified a certain type of engine and he didn't use that kind. That ended all royalties; *nobody* was using that kind of engine. The licensees profited more by losing the suit than they would have by winning it but, regardless of legal quibbles, Selden's position as a pioneer still stands.

In spite of Lenoir, the French at first stuck to steam as motive power. They specialized in outsize road locomotives designed to pull whole trains of coaches on the roads. Amédée Bollée experimented with these things and with an English style steam diligence; then he moved on to build *La Mancelle,* which has been called a steam

The original Hansom cab, 1834

Selden's gasoline automobile, drawn from his original Patent Office model, 1879

victoria but which the French label a *calèche à vapeur.* She stands today as part of the great collection of antique vehicles in the Château de Compiègne.

Bollée used flat steel spokes and steel rims for his wheels to add a little resilience and to absorb the punishment of the rough roads which sometimes knocked wooden wheels to pieces. It took two men to run *La Mancelle,* a steersman and an engineer, but she would carry *papa, maman,* and the small Bollées on a Sunday outing, and she had a weird grace that was hers alone.

On the hundredth anniversary of the Declaration of Independence a Centennial Exposition was held in Philadelphia. The United States proudly showed off its own products and had a chance to see the exhibits of older nations. In quality the coach-

work of such American masters as Brewster stood up well against the European examples but, when it came to swank, the Englishmen were able to pop the Yankees' eyes.

The park drag as exhibited by Messrs. Hooper and Company (approved by the Four-in-Hand Club of London) was used by wealthy English sportsmen to carry whole house parties to the races, to cricket and football matches, or merely across country to a picnic spot. It was excellent for watching sporting events because its height made it a portable grandstand. The "parking" of these vehicles for this purpose has given its name to our most widespread modern nuisance.

Once introduced, the drag rapidly became popular in America among those who could afford it and soon excellent ones were

La Mancelle, French steam calèche, 1878

being built over here. The later examples seem to have been a little squarer in outline and a little shorter than the one below.

Though lighter in construction and more conservative in color, the drag was an obvious descendant of the English stagecoach. As has been mentioned, all such coaches had individual names. One of them, called the *Tally-ho,* was brought to this country. It attracted considerable attention and the public, making no nice distinctions, lumped all coaches together as tallyhos. The drag was also known as a coach-and-four, but so was a road coach. It is necessary to distinguish between the two because the road coach reappeared in the coaching revival around the turn of the century when road coaches were actually put back on scheduled runs both in England and in the United States. These were driven by gentlemen "whips" and were subsidized by wealthy enthusiasts. The vehicles were painted in the traditional bright, contrasting colors and could properly be turned out with unmatched teams.

The horses of the drag had to be inconspicuously colored, as nearly alike as possible, and possessed of some style in the way of gait. Their harness, which was prescribed in every detail, had no breechings. Stopping was accomplished by the wheel horses, which took the strain on the backs of their necks; it was bad form to use the hand brake except on steep hills. The whole turn-out was a formal affair and everything about it was done "correctly." For instance, though a road coach carried its lamps in their brackets at all times, on a park drag they must never be seen except when in use at night.

The front seat of a drag held only two. On the right the driver was propped on a high wedge-shaped cushion, with a light driving apron over his knees to protect his trousers. The privileged passenger beside him was more often than not a lady. She reached her place by means of a folding ladder placed by a groom. The driver got up in accordance with a prescribed but logical system. Grasping a handle just below the seat, he first put his left foot on the hub, then his right foot on the top of the roller bolt to which the off wheeler's trace was hitched (it had a large cap for the purpose), next his left foot was brought up to the front step on the side of the boot, and finally his right foot landed on the floor board.

The two seats on the roof projected far enough over the sides of the body to permit each of them to hold four passengers, the rear four facing backward. On a road coach the back seat also held four, the guard occupying the left-hand end. The

Park drag, about 1880

back seat of the park drag, however, was wide enough for only the two liveried grooms, one of whom was expected to be a virtuoso on the long coach horn.

There isn't enough space here for a detailed account of the technique of driving four horses in approved style. The sketch on page 70 shows how the reins were held in the left hand. On reaching a corner, the coachman drew in the required lead rein with his right hand, making a loop in it which projected above his left hand and was held by his left thumb. This put a strain on the same side of both leaders' mouths and was called "pointing" or "making a point." Once the loop was formed it was necessary to "opposition" the wheelers by putting tension with the right hand on the *opposite* wheel rein, so that they wouldn't swing with the leaders and turn the corner too soon. Inexpert driving on a turn could easily break the pole or throw the wheelers off their feet.

Since the ordinary road cart (see page 62) was already known as a sulky, its lighter brother on the next page probably had the same name after it was introduced on the trotting track about 1870. The racing sulky had the spindling construction of the racing wagon and, of course, was lighter by two wheels, so faster average time was made with it.

Racing harness was also made as light as possible. There was just enough of it to control the animal and pull the cart. The details of this harness and much other harness in this book were taken from *The Harness Makers' Illustrated Manual*, published in 1880.

Except that its elegant decoration was later simplified, the vehicle at the bottom of the next page might stand as the final version of the horse-drawn omnibus. Some may have been a little larger, but no basic improvement appeared before omnibuses went the way of post-chaises. This bus seated sixteen riders on long side seats and carried as many more as could stand in the aisle or cling to the rear steps. Since the pulling was done by only two horses, transit was not rapid. John Stephenson and Company, who built these omnibuses, turned them out for use all over the country; they also were the builders of the first American horsecar.

One of the things that made life exciting for omnibus drivers was the high-wheeled bicycle. In England these things were called "penny-farthings" from the relative size of the two wheels; over here they came to be known as "ordinary" bicycles, a name they would seem to have acquired *after* the invention of the "safety." Anyway, they were made with four-foot or even five-foot front

Wooden racing sulky

wheels and one-foot or even ten-inch back wheels and were equipped with solid rubber tires. Both wheels had staggered wire spokes, a new idea. Instead of standing on stiff wooden spokes upon the lower half of the rim, the hub was suspended from the upper half.

The high bikes had brakes which operated from the handle bar and clamped down on the top of the big wheel, but often speed was reduced by pressing with the sole of one's shoe on the top of the small wheel. There was a little step low on the back of the frame; it was helpful in mounting and a smart rider could stand on it going down a steep hill. Not only could he

brake, or help to brake, with his free foot but also his weight was better placed for an encounter with a small stone in the roadway. When the rider was on the seat, with his weight high and fairly far forward, such a stone could and often did cause a sudden forward somersault involving both man and machine. This accident was called a "header" and, though its precise duplicate is now rare indeed, we have broadened the meaning of the word and kept it useful.

There were professional high-wheel men who raced the things, and they were also quite generally ridden by men and boys. No doubt there were girls who rode them, too, but not usually in public. Very dashing

New York omnibus and "ordinary" bicycle, about 1880

damsels sometimes stood on the little step and clung to the belt of a devoted swain to hitch a short ride, but this was considered no conduct for a lady.

Because taking a header was often serious or even fatal, the "Star" bicycle was developed. This one wore its small wheel out front but, though it was a notably safer machine, it never became very popular. The weight of the Star's rider had to be placed a little forward of the center of its large wheel, hence the pedals couldn't be cranked directly on its axle. Instead of being rotated, they were pumped up and down and did their driving by pulling on spring-returned leather belts which were wrapped on the large hub.

There's a shaky tradition that the victoria originated from a pony phaeton presented to the Queen by a loyal carriage-builder. The presentation was made all right, but Her Majesty's carriage had no seat for a coachman, and victorias had them.

The victoria was evolved by adding a driver's seat to the low George IV phaeton. It seems that such a carriage met with indifference when it was introduced in London quite a while before the presentation mentioned above, but the French liked it, and copied it, and christened it a *milord*. Then

"Star" bicycle, about 1880

the Prince of Wales admired a milord in Paris and took one home. His and his mother's sponsorship quickly popularized the cabriolet, as it was confusingly called.

The victoria has been called the "handsomest vehicle ever built by man" and that may well be true. Its shape still dimly affects the shapes of the cars we drive today; our "fenders" are the descendants of the victoria's "mud guards." Certainly, in days that are gone, a victoria was the very symbol of elegance. A handsome woman; two men in livery on the box; a matched pair of "spanking bays" in silver-mounted harness: that was a heartening spectacle for anybody but a Communist. The sunlight danced on the varnished spokes and glinted on the silver lamps. The rubber-tired wheels rumbled quietly on the Belgian blocks, and

Panel-boot victoria, or cabriolet, about 1885

Lady's drop-front phaeton, about 1885

the padded shoes of the horses hit the paving with a precise, muffled cadence that might stand as the musical motif of the age when security was taken for granted.

The panel boot, sometimes called the "box," was a receptacle for packages under the driver's seat. With flat sides, it was easier to make and better looking than the old pod shape shown in the drawing of the barouche on page 66. Not all victorias had this boot.

The sweeping lines of the victoria came from King George's phaeton, which was only one of a large family, not all of which were so graceful. Many kinds of phaeton were in use in the United States, but a standard variety developed which was only slightly less common than the buggy. This

standard drop-front phaeton's floor was close to the ground, and its collapsible top was exceptionally large and came far forward, giving more than usual weather protection. Because of these things, but for different reasons, the phaeton was favored by ladies and doctors. These carriages weren't all made like the one in the drawing, which is of the simple, popular type driven by the grandmothers or great-grandmothers of most of us. Some phaetons were built to attract attention and were given light-colored parasol tops, or great sweeping dashboards which reached over the horse's back, or rumble seats to carry a footman who could assist her ladyship to the ground and hold the horses while she "left a card" with an acquaintance.

Probably the only American carriage that ever had any popularity in England was the very high, very lightly built spider phaeton. In general the sturdy Briton didn't care for the "flimsy" construction of American carriages.

Karl Benz's gasoline tricycle, 1885

78

In its first years the motorcar had trouble making up its mind whether it wanted to be a carriage or a bicycle; in fact, it's possible to discern traces of both mixed in with the aircraft streamlining of modern cars. Karl Benz, the German, was a bicycle man in spite of the three wheels he put on his machine. His was the first practical gasoline automobile. Lenoir and Marcus had pioneered before him, so had Selden, and Gottfried Daimler turned out a fearsome-looking wooden motorcycle in the same year, 1885; but if the ignition was feeling happy, you could start Benz's machine and get in and go places in it. The horizontal, water-cooled one-cylinder engine was mounted where an engine is most efficiently mounted, in the rear. The wire wheels had solid rubber tires like those of the high-wheeled bicycles. No claims were made for the first Benz as a speedster. Later cars of the same name did better. In 1909 Barney Oldfield set a world's record of 132 miles per hour in the "Blitzen Benz."

After Selden's try, American inventors passed up gasoline for a while, but steam made a good firm step forward. In 1886 Ransom E. Olds built a steam tricycle in Lansing, Michigan. It must have been successful because in 1898 he added one more wheel to it and sold it to a customer in Bombay! That at least made him the first exporter of automobiles from the United States. Ten years later, after the Duryeas

Ransom Olds's steam tricycle, 1887

and Haynes had built their cars, Olds brought out a gasoline model. He went on from it to become a successful motorcar manufacturer, which seems to have been what he was after in the first place. Like Benz and Daimler his name is still connected with automobiles.

Mr. Olds didn't hold with bicycles; even though his front fork was straight from the "Star," he was a carriage man. In spite of its being a tricycle, the general appearance of the machine forecasts the runabout form of most of the early twentieth-century cars, including his own "Merry Oldsmobile."

If anything bears out the popular conception that "the winters ain't as cold as they usta be" it would seem to be the fact that our ancestors rode on runners for all or most of the winter. Everybody had a sleigh and used it; sleigh races were held

Cutter and bobsleigh, about 1885

in areas where nowadays the kids have to wait anxiously for a day or two with snow enough for coasting. Of course, there is snow enough up north, but how much exercise would a sleigh get in, say, New York City in any of our recent winters? One old diary records the last of forty consecutive days of sleighing there.

The cutter or "one-horse open sleigh" of the song is familiar to most of us. We have seen it on Christmas cards if nowhere else. Those who have ridden in one behind a good horse and a string of bells have known a certain satisfaction that is not to be had in any other land vehicle. Many a swain took his girl for a sleigh ride and dumped her into the snow. Turning a cutter too sharply or pulling a runner out of a frozen rut could easily upset it. This wasn't nearly so likely to happen with a bobsleigh, since the short runners of the bobs and the fact that the front one was movable made steering simple. Bobsleighs were used for winter hauling and sometimes were created by replacing the wheels of a wagon with temporary runners.

That sleighs were also used as more formal carriages and were sometimes very handsomely finished is not so well known. Some were enclosed like broughams and were known as "booby huts." A good sleigh was built as carefully as a carriage and was given the same elaborate paint job. Before the days of spray guns coach-painting was a long, exacting process, but the result produced a rich depth of color and a glowing beauty of finish which is no longer to be seen. In the nineteenth century a fine carriage was given as many as eight undercoats, each rubbed to smoothness, before the final color was put on, and *it* was given not less than two coats. Over the last color six coats of copal varnish were applied, all of them except the last being rubbed. The final finish to the last coat was not given until the carriage had been in use a month. It was then rubbed with soft rottenstone and polished with flour and oil. After that it was only necessary to wash it *every day* with cold water. Lines of color were frequently applied as ornament to both the body and the undercarriage. The work was done *freehand* with a very long-haired little brush called a striper.

Perhaps one reason New Yorkers would have none of the hansom cab when it was first introduced was that they were happy with what they had. Here you see what that was: the Gurney cab with its tin-box body, and the sedan cab which was fancied as looking like a sedan chair. Passengers entered both cabs from the rear. The Gurney had side seats for four, facing each other. The sedan carried only two passengers who entered and closed the door; the driver then lowered the two hinged sections of their seat from up front, with a crank. The construction of both of these vehicles followed the theory of the early hansom, putting the weight of the passengers behind the axle, balanced by the driver's weight ahead of it.

Gurney cab and sedan cab, about 1885

Steam fire engine, 1890

To small boys this was the most exciting vehicle ever built. For all the roar and the screaming siren, no motor-driven engine ever seems to go as fast as this kind did behind three galloping horses. Surely too, no modern firehouse is so fascinating as the one that housed the steam-pumper which stood glittering in the middle of the floor with the harness for its steeds slung from the ceiling before it. The horses were stabled around the engine in stalls which were open at both ends except for light chains across their fronts.

When the alarm sounded, the chains dropped and each horse went at once to his proper place. Some men came running and others slid down the brass poles from the dormitory above. Down came the harness, slapping the horses as it landed and, perfectly trained though they were and practicing every day, they never learned not to jump a little when it hit them. Each fireman had his own straps to attend to and he made quick work of them. The harness was much simplified for quick handling; snap hooks largely replacing buckles.

The grate was poked up to raise the steam pressure, the big doors banged open, the driver gathered up his "ribbons," the engineer jumped aboard the rear platform, the gong clanged, iron-shod hoofs struck sparks from the stone floor, and the juggernaut, belching smoke, went pounding off down the street—a sight of sheer grandeur!

At almost the other extreme of speed was the Studebaker farm wagon, which went to no fires and seldom moved faster than a lumbering jog. Some such wagon, drawn by a heavy team, was on every American farm. More than likely it was built by the Studebaker "works" or by the local blacksmith. Though better finished and with far better iron fittings, the running gear of this wagon was not essentially different from that of the Roman wagon on page 16. Many a one of them with its long tongue (pole) cut short now rolls ignominiously behind a tractor. A note of propriety—it is correct to speak of a pair of farm horses as a "team"; with all other horses a team is three or more.

Some city drays had springs, but farm

Studebaker farm wagon, 1890

Duryea gasoline phaeton, 1893

wagons were made "dead axle," the only springs being a couple of small ones under the removable seat. Standing on the bed of such a wagon as it rumbled along a macadam road would jar the teeth of a youngster and make the soles of his feet tingle in an entirely delightful way.

Charles Duryea designed the so-called gasoline buggy above; his brother Frank built and drove it. Frank contributed the ideas for the make-and-break ignition and the jet carburetor. It made its first run at Springfield, Massachusetts, in September, 1893. The brothers had made a try a year earlier but their engine had not been powerful enough. Their machine is unquestioned as the first successful American gasoline automobile. It was the original "horseless carriage."

The Duryeas pretty obviously built their car around a standard horse phaeton. At first it had friction drive; later this was changed to a more satisfactory gear transmission, which is still in place. The old vehicle has now lost its drive chains and its lamps; its dashboard is tattered and its top has sagged; otherwise, it stands in the National Museum just as the brothers built it.

The one-cylinder, four-horsepower engine is hung on the rear axle. Water circulated around the cylinder and returned to a cooling tank at one side. A train of gears transferred power from the motor to a horizontal jack shaft, which ran straight across the car not far above the road. From pinions on the ends of this shaft, chains drove large sprockets bolted to the spokes of the rear wheels.

Gears were shifted by raising and lowering the steering tiller; this gave two speeds forward and a reverse. Carriages steered by swinging the whole front axle. The Duryea front axle was fixed, and each front wheel was pivoted separately, like the wheels of a modern car. The wheels themselves, however, were the original wood-spoked, iron-tired ones that came on the phaeton.

By 1896 the Duryea brothers were in the automobile manufacturing business. Thirteen improved cars were built. They were all alike, with the mechanism tidied up and put under cover and with pneumatic tires on their wooden "artillery" wheels. They were all sold to the public, and one of them was an attraction in Barnum and Bailey's circus.

In spite of "stink-wagon" experiments, France was still loyal to steam and by 1890 Serpollet had a light, workable, coiled-tube boiler and was able to bring out a dashing little steamer which had no smokestack whatever. As we've seen, Ransom Olds had done the same thing earlier, but he didn't follow it up. He switched to gasoline instead.

Serpollet steamer, 1890

Though the safety bicycle returned to the wheel arrangement of the velocipede, its rear wheel was for some time made slightly smaller than the front one, as a kind of memorial to the ordinary. It was in every other way radically new. Nothing like its light, tubular frame had been seen before; its pneumatic tires were an innovation (the earliest safeties had solid tires); but, above all, the sprocket-and-chain which carried power from the pedals to the *back* wheel and caused that wheel to turn faster than the pedals did was an entirely novel idea. As a variation, some bikes were driven by a shaft and bevel gears. The coaster-brake, which allows freewheeling and brakes by reverse pedaling, was invented later. The brakes on the old safeties pressed the tire of the front wheel when one squeezed a lever on the handle bar.

Probably the feminine drop frame for

the safety bicycle should be credited with releasing American girls from the rigid code which made them act like "ladies" and getting them outdoors and into active sports. By eliminating the top bar at some sacrifice of strength, it allowed a girl in a longish skirt to ride with no serious loss of dignity or modesty. Divided skirts or bloomers came to be "the thing," however.

It is hard for us to understand how wonderful the bicycle seemed to the nineties. Here was rapid transportation within the reach of all. Cycling for both sexes became a craze never equaled in any sport before or since. Cycle clubs were formed everywhere. Races were a dime a dozen. Strong men proved their hardihood by riding a "century," a hundred miles in one day. Large towns set aside roads to be used exclusively by cyclists. In one city the wide path encircled a reservoir, and on every

"Safety" bicycles, about 1895

Buckboard, about 1895

Governess cart, about 1895

clear Sunday it was solidy packed with cyclists of all ages pedaling gravely round and round. Cycle policemen moved with the crowd and dealt with "mashers" and "scorchers."

Since there was no kind of clutch on the safety, coasting downhill presented difficulties. The pedals fairly flew around. Letting one's feet rotate with them was both uncomfortable and undignified. To avoid this, the safety had a metal spur on each side of the front-wheel fork; on these the feet could be rested while coasting. To get back on to the spinning pedals with aplomb was a trick which required practice.

Bicycles are still popular. Many more are sold every year now than at the height of the craze; even set against the present population, that's a lot of "wheels." In this country most of them are ridden for sport, but in Europe, especially in the Scandi-

navian countries, bicycles carry a large proportion of the traveling public.

Though it had no springs whatever, the buckboard above wasn't as rough-riding as it would appear to be. The reason lay in the resilience of the long strips which formed its floor. Originally the floor was a single, very wide board (hence the name), but slats were substituted when such wide lumber became hard to find. The buckboard is an original American idea; its origins go back to colonial days. In the early years of the twentieth century some buckboards were handsome vehicles. Usually they were not painted; the natural wood was varnished.

No country estate was complete without a governess cart. Drawn by a pony—a fat one was best—it amused the kids, aired the baby, and was sent to the village for five pounds of sugar or over to Mr. Bannister's

The first Ford and a rural mail wagon, 1896

with a custard for his ailing lady. Driving it over the right shoulder from its side seat wasn't easy on the back, especially as the pony's short stride made a continual joggling. Not all governess carts had basket bodies, but most of them did.

The descendants of the car on the preceding page number in the millions. Henry Ford himself built it. This one's running-gear was assembled with no idea of producing a useful vehicle but simply to test the performance of the engine. Ford had already built a little, one-cylinder engine on his wife's kitchen table. By 1896 he had progressed to the model which he put into the little car. It had two parallel cylinders placed on either side of a large flywheel, and it developed four horsepower. There was no crankcase; the connecting rods did their work out in the open air.

The car itself was little more than a rectangular frame mounted on four bicycle wheels and supporting a seat. Ford called it a quadricycle. It did twenty miles an hour, which was high speed for its day and plenty fast enough to go over rough roads on solid rubber tires and a springless rear axle. The front axle rode on small elliptical springs. The engine was mounted in back and was belted to a jackshaft under the front of the seat. Going forward there was a low and a high speed, but to back up you had to get out and push. This wasn't too hard to do because the whole machine weighed only five hundred pounds. Steering was accomplished with the customary tiller, with its handle over the middle of the seat; steering wheels didn't appear on any American car before 1901.

Ford's quadricycle had little about it that someone else hadn't previously done as well. Mechanics all over the country were beginning to tinker with four wheels and an engine. The Ford machine would probably have been forgotten along with many

others except for what its inventor did to automobiles afterwards. He seems to have had the production of a cheap car as his goal almost from his first motor experiments.

The Duryeas and Ford hung their engines on the rear axle. The French, who had been building cars to sell since 1890, usually put theirs under the seat. Since these primitive putt-putts were at best temperamental and sometimes blew up, this location tended to make passengers ill at ease. In '92 René Panhard decided to relieve the nervous tension, so he moved his Daimler motor clear up front, putting it under a "bonnet" as far from the wheels he was driving as he could get it. In the course of the move he invented the sliding-gear transmission which, somewhat elaborated, is still with us.

Most of the American cars copied the under-the-seat position for the engine in their first years of commercial production; then, starting with Locomobile in 1902, they all blindly followed Panhard. Though engineers have long agreed that weight in the rear and a short drive shaft are the ideal arrangement, the public still stiffly resists any attempt to change to it.

The tandem cart on the next page was the final version of the cocking cart (page 61), and to drive one was about as smart as you could get at the turn of the century. Such a turn-out was roughly equivalent to the Continental sports car of today: possibly a little flashy but the last word. High-wheeled gigs without a seat for a servant

Panhard gasoline carriage, 1892

were also used for tandems. Incidentally, when a lady climbed into a high cart, where her long skirt might brush the wheel and be soiled, a curved basketwork guard was put over the tire.

It would be entirely improper to hitch a tandem to a buggy. When tandems are exhibited at modern horse shows they are almost always driven to a stanhope gig, which is too low to give the full dashing effect that could be obtained with the rather dangerously high cocking cart. Handling a tandem is no sport for a novice. The driving procedure is the same as for a four-in-hand but perhaps more difficult because the leader is placed so far ahead. If he is not kept constantly where he belongs, the smart appearance of the outfit quickly vanishes in a welter of legs and tangled traces.

A pair to be driven this way must be trained to the job and should be schooled from colthood to step "high, wide, and handsome." The shaft horse is always larger and heavier than the leader and wears correspondingly heavier harness. No breeching is used. On the shafts, just behind the tugs which support them, there are projecting metal stops. The horse can hold back against these with the tugs, taking the weight of the carriage on his back rather than on his thighs. It is usual for the shaft horse to wear a kicking strap, which passes from shaft to shaft over his hips and serves to check any emotional impulse he might have to kick the front out of the cart.

Even in 1900 a high-wheeled tandem cart was not to be met in every street as were the vehicles at the other end of the social scale, the squadrons of lowly wagons which served the needs of cities. They came in all sizes and shapes. There were milk wagons and pie wagons and lunch wagons, each with a body of special shape. There was also the lumber wagon, which had no body at all but which could be lengthened or shortened to accommodate its load by moving the rear truck on the long perch.

On the next page are two familiar in every urban neighborhood up to a comparatively short time ago: the hooded ice wagon with its low rear step to tempt small fry to hitch a ride, and its cool, dank interior from which ice chips could be snitched on hot days; and the coal wagon with its inevitable umbrella and its box body which could be raised and tilted to allow the load to run down an iron chute with a speech-blanketing roar.

The thing directly opposite shows the influence of the hansom cab on the motorcar.

Ice wagon, about 1900

Coal wagon, about 1900

Following it in the illustration is the hansom itself in its final American form. Beyond the electric is a large dead-axle (springless) dray which did the sort of heavy hauling for which a flat-bodied truck would be used now. Around 1900 there was great interest in electric automobiles, and much was expected of their development. The electric hansom, among its other distinctions, was steered by its back wheels, which were smaller than the front ones.

Such contraptions actually operated on the streets of New York from 1898 to as late maybe as 1904. That would be about the year that a timorous maiden aunt in charge of a small boy and a cocker spaniel was somehow betrayed, or perhaps cajoled, into hiring one to cross the city. It was a fear-

some experience. For years afterward a hunted look came into the lady's eye whenever the ride was mentioned.

You may wonder why, once the automobile has been introduced, this book continues to concern itself with carriages and wagons. It's true that there was almost no change in their design after 1900, but it's also true that they were in wide use for a long time after that, in fact, they predominated until 1913 and were given a new boost five years later by the wartime shortage of gasoline. It's true, too, that within a hundred miles of where you are reading anywhere in the United States there are respectable, prosperous Americans who still drive to town behind Dobbin.

As the buggy was the standard carriage

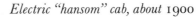

Electric "hansom" cab, about 1900

Fringed-top surrey, about 1900

for two passengers in this country, so the surrey was for four. Most surreys had rigid, fringed tops like the one in the illustration, but not all; some had "auto" tops, some had "parasol" tops, and some had no tops whatever. The auto top folded like a buggy top. It was attached to the arms of both seats and, on the rare occasions when it was put down, it had to be detached at the front. The parasol or canopy top was actually built like a rectangular parasol and was supported by a single iron stanchion fixed against the back of the front seat. Its four corners were guyed to the body by cords. A silly looking thing, it was usually cream-colored or white and sported a thick, six-inch-deep fringe around its edge.

Occasionally you would see a lone horse laboring to pull a surrey with a whole family in it but more often the surrey was hitched behind a pair. "Drop" poles, which were branched so they could be clipped onto the axletree like shafts, were quite common. These were always used with a "yoke,"

which slipped over the pole tip and had its ends attached to the bottoms of the horses' collars. It served to keep the horses apart, and they could hold the carriage back with it.

About this time some people began to assert that "the automobile is here to stay," while others yelled "get a horse!" The illustration shows the "Merry Oldsmobile" that the song was written about. The Oldsmobile was called a "one-lunger" because it clung to a single-cylinder engine. You could buy one for $650 and, if you remembered to keep the oil cups filled and didn't mind stopping now and then to put water in the radiator, it would usually get you where you were going.

Like the French diligence of 1771 the 1902 Olds had its body balanced on long springs which connected the axles. The engine was under the seat and was started by cranking from the left side of the car. The juice to spark the ignition came from a bundle of dry batteries, their voltage stepped up by a vibrator coil. When the batteries were dead you'd had it, unless you carried spares.

The single brake band that gripped the drive shaft would stop the machine if you gave it time enough. The three-inch tires were pumped up hard with some sixty pounds of air, and they were pumped up by hand! There were no free-air hoses.

As cars became powerful enough to carry

One-cylinder gasoline Oldsmobile, 1902

more than two passengers, a seating problem arose. It was solved in many ways, among them the magnificent *tonneau*. This was usually higher than the front seat and was entered by a narrow door in the middle of the rear. By 1905 some makers were advertising side-entrance tonneaus. A top was an extra at a hundred dollars. A windshield was extra, too. In addition to the dependable oil lamps, most cars by 1904 had carbide gas headlights. The light they gave was brilliantly white and they caught with a fine POP when a match was touched to them. The gas was generated in a small tower which stood on a running board—and a thorough mess it was.

There are still those who bemoan the disappearance of the steam car. The bearded Stanley twins did much to perfect it with their "Flying Teakettle." The White was built under a Stanley license. A steamer needed no clutch and no gears, only a brake and a hand throttle. If you wanted power for a hill you simply opened the throttle wider and went up. As to speed, the throttle gave it to you; the model in the drawing did forty. Later steamers, especially the Stanley, were really fast.

By 1904 most cars had abandoned the tiller for the steering wheel. Fenders were merely mudguards. Bumpers? No. You weren't supposed to run into things, and there were few things to run into you. The

White steamer, 1904

horn was a brass bugle affair, blown by squeezing a rubber bulb. This produced the traditional HONK!

The rockaway, like the buggy, was a 100% American carriage. It came in a wide variety of styles, all of which were called depot wagons, station wagons, Dayton wagons, or Pittsburg cut-unders, not for the purpose of distinguishing one kind from another, but simply as a matter of taste. The family name is said to have resulted from a lie told by a carriage dealer to conceal his source of supply, which was not Rockaway. The motorized descendants of this vehicle are called station wagons, estate wagons, and other things.

Though it has been claimed that the first rockaway was built at Jamaica, Long Island, in 1830, the neighborhood of Philadelphia seems a likelier birthplace for it. Coachees had been very popular there (see page 48). The Germantown wagon of the area was the legitimate child of the

Light curtain rockaway, about 1905

coachee, and the Germantown minus its two front top-supports was almost a rock-away.

Paul Downing has described five distinct types of rockaway, ranging from simple, curtained carryalls like the one in the illustration to glazed carriages which looked like broughams, except that they kept the extended top which sheltered the driver. This democratic idea was borrowed by the surrey, and Colonel Downing points out that early limousine bodies on both sides of the Atlantic were direct copies of the coupé rockaway.

Most rockaways were painted but late in the nineteenth century some of them began to be finished in natural wood, varnished like the one illustrated. These were built with the framing on the outside of the body as the old pleasure wagon had been. Framing and varnish together were responsible for the paneling of the gasoline station wagon and for the (now fading) custom of giving it a wooden body or the simulation of one.

Like the last rockaways, the dog cart was usually a country vehicle. It was originally designed in England for the use of hunters. Its two seats, back to back, were set over a ventilated box in which bird dogs could travel. It was found that the box could also carry groceries or laundry so, because they were handy and sporty, dog carts came to be driven as utility carts by people with no interest whatever in hunting. Most dog carts had a crank within the driver's reach,

which moved the body on the gear to balance the load.

Once a youngster and his grandmother were left alone on the rear seat of a bright yellow dog cart, backed up to a village curb. Somewhere up front a strap broke. The shafts swung upward; Sonny clawing the air and Grandma in a fury of starched petticoats and shattered dignity slid to the pavement with no recorded injuries. The horse had apparently been expecting the worst, for he simply stood still.

The Model "T" on the opposite page is the original "Tin Lizzie." Born in 1908, she flourished nineteen years; fifteen million copies of her were made, the last different, but not too different, from the first.

Lizzie was as simple as a car could reasonably be made, with about one third as many parts as it takes to build a car now; but it was the method by which her parts were fabricated and put together that made history. The Model "T" was the first article ever mass-produced. Each worker did one job over and over; his work was brought to him and was moved away on a schedule, based on the time required for each operation. At the end of the production line the cars rolled out of the plant under their own power. Once 9,109 of them rolled out in one day!

Ford made a car without frills, cheap enough for every man to buy. The first one was a sensation at $850. Mass production made them cheaper and cheaper. At one time, though Ford was paying the highest

Dog cart, about 1905

Model "T" Ford, 1908

wages in the industry, you could buy a Model "T" roadster for $290!

Lizzie had her steering wheel on the left side, and before she was through she had persuaded all other American cars to follow her lead. She had no battery; you spun the crank fast enough to get a starting spark from the magneto. Later on starters were added as an extra. There was no foot accelerator; speed was controlled by a hand throttle under the steering wheel. The spark control was there too. It was retarded to give the engine better pull on hills, and it *had* to be retarded when the motor was cranked to avoid a back-kick which would break an arm. When there was nobody to help, you needed agility to get around from the front and shove the spark up before the engine quit.

Gears were shifted with the feet. Reverse had a pedal to itself and was in the middle. For low you shoved the left-hand pedal down to the floor and held it there; letting it clear out gave you high. The third pedal on the right was the foot brake, and there

was a hand brake too. The first model had two hand levers, but one was eliminated later.

If someone bought a 1908 Maxwell new, he paid $1,750 for it. This one ran on magneto too but started on batteries. After you got her going by cranking, you quickly threw the switch across from the side marked BAT to the side marked MAG. Gear-shifting was "progressive." There were three speeds forward, but to get from low to high or back again you *had* to go through second. This Maxwell used carbide headlights, but tanks of compressed gas called Prestolite were available by this time. They were less nuisance and quite satisfactory as long as you always carried a spare tank.

Electrics were the first closed cars; some of them were glassed-in as early as 1900. Once it was believed by most people that the motorcar of the future would certainly be electric. A genius was to invent a super-battery, light, compact, and powerful, with which it would be possible to make five-hundred-mile trips without recharging. Or,

Maxwell touring car, 1908

even better, power would be received by "wireless" and paid for by taxes. Well, who knows? Maybe that's not so funny.

The electrics got off to a good start because a usable motor had already been developed which was lighter than any known steam engine and far more dependable than any of the early gasoline devices. At first a couple of electric motors were mounted right on the rear axle. Thirty or forty storage batteries went into the front and rear compartments and were charged in the garage overnight. At its best the electric traveled about a hundred miles on one charging but not at its wasteful top speed of twenty miles an hour.

Because they were slow, quiet, and very simple to operate, electrics appealed to the kind of ladies who wanted to drive themselves but hadn't the sporting instincts required for dealing with an early gasoline or steam car. Doctors found buggy-top electrics convenient for the frequent stops required by house calls; electrics didn't have to be cranked.

Even quite recently it wasn't too unusual to see a ton or so of glazed conservatory whirring sedately along a city street. Elderly folk gave up their electrics reluctantly, but they weren't agile enough for modern traffic. However, not all electrics were slow; in very early days some were geared up and used for racing.

Most electrics were chain-driven and so were practically all early commercial cars, regardless of motive power. A panel truck at first was called a delivery car. The word truck seems to have been applied at first only to heavy-duty, box-bodied affairs, such as might be used on a farm or by a contractor. The 1911 Autocar was gasoline-powered and, since it was widely used by department stores for making deliveries, it may be taken as an indication of increased dependability in the internal-combustion engine. Because of dependability and also because of frequent stops, many early delivery cars were electrics. The earliest commercial truck of all was built in 1892 for a French department store. It was a steamer. It had little space in it for anything except its engine and boiler, but its advertising value left nothing to be desired.

Stylish persons who formerly had ridden about in closed carriages and victorias graduated naturally to limousines and landaulets. This was expensive. A 1912 Packard landaulet could set you back $6,500. Such a car demanded a chauffeur; there may have been somewhere an eccentric gentleman who would be seen driving his own landaulet, but he would have been thought very eccentric indeed.

Even landaulets were cranked. The 1911 Cadillac had come out with a self-starter, and electric lights had appeared, though, curiously enough, some high-grade cars clung to Prestolite for several years. When the Model "T" Ford changed over, the lights were naturally powered by the magneto, and their brilliance fluctuated disconcertingly with the speed of the motor.

The "sport model" is the natural child of the racing car. Perhaps the first American sports car was the Apperson Jackrabbit of 1907 but, since it sold for $5,000, it can't be said ever to have become popular. It was the Stutz Bearcat that gave his sense of superiority to Joe College. "Rakish" was a favorite word with the advertising writers of this time, and rakish was the word for the Bearcat. Painted red, with a leather strap over its hood, with an oval gas tank back of its bucket seat, and with two spare tires behind the tank, it produced an effect that was breath-stopping to anybody under thirty. The radiator ornament was a Motometer which registered water temperature and which could be read from the driver's seat by a sharp eye.

Front doors as well as rear ones became standard by 1914. Chevrolet made a cheap car which, unlike the Ford, was changed from year to year to follow style trends. It sold so well that it finally forced Ford to abandon the Model "T." The Baby Grand sedan was equipped with a crank and Prestolite lamps, but for $125 extra a self-starter and electric lights were provided.

Bumpers; disk wheels: these were not entirely new, but mechanical four-wheel brakes were, and so was the straight line along the top of the Chrysler's hood and body. It's hard now to remember how really slick that looked when you saw it for the first

Stutz Bearcat roadster, 1914

time—slick but startling. Though people like to own the latest thing, most of them hesitate to try anything so radically new that it might spotlight them and make them appear ridiculous. So they held off from the first Chrysler and it wasn't a howling financial success in spite of the furor it raised; but it influenced the design of all American cars made after it. That influence can be seen in the new designs for 1926, and next year's new models will show traces of lines they inherited from the 1924 Chrysler.

When, by buying more Chevrolets than Fords, the public gave unmistakable notice that it valued stylishness above cheap transportation, Ford had to get into line or quit. It meant a reversal of cherished policy which was heartbreaking to Henry Ford, and it also meant a huge retooling job, for which the plant was closed down for nearly six months. Enormous and costly steel dies had to be shaped for stamping out body parts. Many series of elaborate "jigs" had to be designed and made for holding mechanical parts and guiding the operations performed on them.

In December, 1928, the Model "A" appeared and, though it was not radical except as a Ford, it stirred up as much excitement as the Chrysler had created. It

Chrysler roadster, 1924

American double-decker bus, 1932

was neat on its wire wheels, equipped with four-wheel brakes. It had a hand-lever gearshift and a shatterproof windshield. This last turned rapidly opaque and had to be replaced on all the earlier cars.

The Model "A" was real quality goods, one of the simplest and sturdiest automobiles ever built by anybody. The roadster sold for $385, the four-door sedan for $570.

Open-top double-deckers like the one illustrated, and others with the upper deck roofed over, were used for years in New York and in other American cities, but it was finally found that the gain in passenger capacity didn't offset the confusion and loss of time caused by getting people upstairs and down again. The stairs had to be too narrow for passing. In clear weather the top of a bus was a pleasant place to ride, and there were howls when the two-story ones were abolished.

The first highway buses were simply covered trucks with seats and windows. Most of them had solid tires. They jolted back and forth between country towns for some years and their routes were gradually lengthened until, by 1930, with improved carriers, intercity bus lines were offering real competition to the railroads.

The road bus of 1930 looked as much as possible like an extended sedan. The windows were often curtained in the belief that a luxurious touch was being added. But this bus was built from the road up to be a bus. It rode on pneumatic tires and was far faster and obviously more comfortable than the old truck-buses. However, people who rode over the rear wheels sat sideways as in the boote of an ancient coach; often they rested their feet on a narrow shelf.

The "Airflow" was another Chrysler that was a failure with the public. It introduced streamlining, but did it so suddenly and so completely that it "looked funny." Nevertheless, it interested designers, and once it had appeared all cars became more and more fish-shaped whether they needed to be or not.

Once the United States was in World War II, manufacture of civilian automobiles virtually ceased, and new cars on hand were rationed out to people with proven need who were engaged in war work or

Highway bus, 1930

Chrysler "Airflow" sedan, 1934

public service. Ordinary folks who owned cars didn't get much use from them, since gas and tires were strictly rationed.

Among many things made by the automobile manufacturers was the Jeep, which the whole world associates with World War II and with the American Army. It was high enough to drive over a rutted road without dragging its belly, short enough to handle easily, and strong enough, with its power delivered to all four wheels, to pull itself out of nearly anything. It was sturdy and ugly, but oddly attractive; it was beloved of all ranks, though riding one at speed over rough country would all but shake a man's soul loose from his body.

Its obvious merits have recommended it for civilian life, and the jeep is car, tractor, and power plant on many a small farm. Unless we are very close to such things, we are apt to think that a Jeep is a Jeep and that its shape never changes. Not so. The Jeep has been modified several times, and the newer ones look quite different from the original model. The latest version, the Bobcat, is nearly two feet shorter than the old Jeep.

After World War II most manufacturers of automobiles got off to the quickest possible start and met the enormous civilian demand by temporarily reviving their pre-war models. Studebaker was the first to bring out an entirely new car; it embodied some features which were radical at the time but which soon became more or less standard in all cars. The great basic change was bringing the sides of the body out to

Military jeep, 1941

the full width of the fenders, thus greatly increasing the seating space and simplifying the outer surface of the car, but not simplifying such a practical matter as replacing a crumpled fender.

The hood and the afterdeck of the new Studebaker were made almost as duplicates of one another, so were the windshield and the back window. It is literally true that people were sometimes confused as to which way the car was headed. Many funny jokes were made about this; but the turret-on-a-deck look of this car is still to be seen in most American automobiles. The only real addition that has been made to it is the preposterous conversion of the rear fenders into tail fins.

The popularity of imported sports cars has led some American manufacturers to bring out high-priced convertibles based upon them and featuring similar breaking-wave front fenders. This shape is not new in Europe. Not long ago there were cars

Studebaker coupe, 1947

over there which weren't too different from the one in the illustration below, and the beginnings of a similar line can be seen in French cars as far back as 1930.

Not many people's tastes or bank balances permit them to drive one, but these extreme cars are interesting for the effect they may have on the kind of automobiles which many people *will* drive. One or two standard makes have already shown up with a wave-shaped chromium strip stuck on the side of the familiar "bug" body, of which it has been said that it looks as if it had been dropped before it was hardened.

The future is for guessing. Seen against the long perspective of man's conveyances the self-propelled vehicle looks like a promising experiment. Far more efficient power plants seem a possibility, mechanical limiting of speed may be a necessity, something must stop the maiming and the slaughter. The craze for display is likely to continue, since it has been with us from the days of the chariot, but it may be expected in time to find other expression than in faltering imitation of aircraft and the juke box.

English sports convertible and American transcontinental bus, 1955